THE STOLEN HISTORY
OF PALAEONTOLOGY

a descent into the deep time of science

Michael Maisch

To my daughters, Lirazel and Laurelin. May they never stop asking me questions.

History is a pack of lies about events that never happened told by people who weren't there. Georges Santayana

CONTENTS

PREFACE

Palaeontologists, both amateurs and professionals, are naturally inclined to follow the news, the latest discoveries and the most outstanding new methods employed to unveil the secrets of the past. And it has never been easier to do so. Information is all over the place, many scientific journals are practically only available online nowadays, and fortunately quite a lot of them have adopted an 'open access' policy, which allows potential readers all over the world to access new publications with no restrictions.

But palaeontology is an ancient science. If mentioned in school at all, it is usually declared to be a science that emerged in the nineteenth century, and most credit for the establishment of the 'new science' is given to Georges Cuvier, the great French (although he was technically born in Mömpelgard, Württemberg) palaeontologist and comparative anatomist, who laid the foundations for generations to come up to the present day.

This is less than half of the story, though. Cuvier could base his studies on the work of generations of scientists who came before him, who excavated and described fossils from all over the world and put them in museums even before he was born. It is with these endeavours of early palaeontologists that we will deal here.

Starting roughly with the year 1758 – because the publication of the 10th edition of Carl von Linné's 'Systema Naturae' and the establishment of modern taxonomy in that year was a game-changer in both palaeontology and biology – we will follow the development of palaeontology through the eighteenth century up to the year 1800, when Cuvier started to dominate the subject like no other before or after him, with some glimpses into times before and after.

Instead of providing an exhaustive – and exhausting – treatise, we will focus on some aspects that have gained particular prominence in the historiography of vertebrate palaeontology.

The stories behind those early researchers and their findings are fascinating and sometimes outright surprising. To me, at least, it was astounding to see what was already known at such an early time, and how scientists now long forgotten except by a few borderline autistic palaeontologists and historians of science, paved the way for Cuvier and everybody who followed in his footsteps.

This book will focus exclusively on vertebrate palaeontology, for two reasons. First and foremost, although I have occasionally worked on invertebrates, I am mainly a vertebrate palaeontologist myself, and I have devoted most of my adult life to the study of fossil vertebrates, so I consider myself as halfway competent. Secondly, they are objectively the most interesting fossils. Not only are they our close relatives, including our remote ancestors, but they are also much more spectacular than the occasional fossil snail or mussel that you can pick up basically everywhere. And they are rare and valuable, and therefore attracted the interest of a large variety of people in the past and present.

The early days of vertebrate palaeontology are also closely tied into the convoluted history of Europe and the Americas during this period (no research was going on in any other continent at the time, at least none that we are aware of), and it is a main focus of this book to take the history of vertebrate palaeontology as a means to maybe elucidate some dark spots in the annals of our own collective 'modern Western culture'.

The late eighteenth and early nineteenth centuries are an extremely important and outright weird and spooky time period. Those were the days when our 'modern world' was established, and just how that happened within such a restricted time frame is a subject that has always fascinated me greatly.

So let us go back in time and have a closer look at the very early days of vertebrate palaeontology. We will meet strange characters, some famous figures from history, scientists long (and undeservedly) forgotten, and a lot of mysteries. And, most importantly, we will see how, and speculate on why so much of that history was distorted, memory-holed and outright stolen.

CHAPTER 1

Leonardo's whale and the
deep time of science

E arly palaeontological research is a black box. The few works from before the nineteenth century that are still in existence have been seen by few and studied by even less. In the age of 'publish or perish', you are, as a young and aspiring palaeontologist, hard–pressed to follow the recent advances in your field. Being flooded with hundreds of publications from all over the world every year, you are certainly not encouraged to research the roots of your science.

You basically accept that it all started with Cuvier (whom you probably never read as well) and carry on. Only a minority of the old works are written in English, the 'lingua franca' of modern science, at least since the end of the second world war in 1945. Instead, the majority of them are in languages exotic and unknown to most current researchers: German, Latin, French, Italian, Dutch, Spanish. And furthermore, as languages change over time, it is rather ancient versions of these languages, so that even with moderate skills in modern German or French, one may find a hard time understanding the contents of these works in full. One may even struggle as a native speaker.

And as these early studies are considered 'not relevant' any more today, they are usually brushed aside. It has become common practice in modern science – out of neglect more than out of necessity – to only cite the most recent works on a certain subject. Maybe the original work in which a species under consideration was first named, but that is about it. Everything before 1900 tends to be widely ignored. Everything before 1800 and the rise of 'modern palaeontology' in the shape of Georges Cuvier and his contemporaries is treated as if it never existed.

It is therefore unsurprising that even experts on certain fossil groups usually don't have a complete idea of the research history of their favourite animals. To give just an example: in their 2007 overview on the fossil history of whales, the authors (Bianucci & Landini 2007) mention and re–figure the well–known illustration of a partial mandible (lower jaw) with three teeth still in place by the early Italian naturalist Agostino Scilla, first published in 1670 in his work *"La vana speculazione disingannata dal senso"*.

The illustration is of such excellent quality, that any vertebrate palaeontologist worth five cents should be immediately able to recognize it as belonging to a specimen of a squalodontid, a family of early predatory toothed whales that was common in the Tertiary. The bone was found on the Island of Malta in strata very probably of Miocene age. Malta is rich in fossils from this time, both vertebrate and invertebrate, and it figures prominently in many of the very early works on palaeontology. Then they go on to tell the reader: "Nevertheless, serious studies on fossil cetaceans began only about 1800. One of the first detailed descriptions appeared in 1824 – the *Recherches sur les ossemens fossiles* by Georges Cuvier." (Bianucci & Landini 2007, p. 35).

This is the usual statement that one can find in literally hundreds of texts on the history of vertebrate palaeontology. One or two examples of early research from the seventeenth or eighteenth century are cited (usually those in which the researchers spectacularly 'failed' to identify their findings), only to come to the conclusion that 'it all really started with Cuvier', who usually also was the first one to give a correct identification for the fossils affected by the mishaps of his borderline retarded predecessors.

To cite Brian Hall (2003), an immensely influential Canadian evolutionary biologist, who, regarding pre-Cuvierian palaeontology, only expresses some avuncular sympathy to Conrad Gesner's work from the sixteenth century: "Georges Cuvier is rightly regarded as 'a palaeontologist, perhaps the first to deserve the name' (Coleman 1964, p. 114)." No. Not true. There were many before him, and they were equally capable. "However, palaeontology existed before there were palaeontologists." Yes. And palaeontologists existed before Cuvier as well. "The origins of palaeontology as a science may be traced to the confirmation by Cuvier in the early part of the nineteenth century, that impressions seen in stones were traces of organisms that lived in the past, species removed from the earth by extinction." No. Not true. None of these things were first explicitly formulated by Cuvier. "The fascination among the public in the United Kingdom for fossils can be traced to an amateur, although a very professional 'amateur', Mary Anning of Lyme Regis, who discovered the first British ichthyosaur in 1812 when she was 12." No. Not true. She didn't. This is just one of innumerable examples of the usual babble you read in peer-reviewed essays by world-renowned scientists in high-ranking journals.

But let us get back to the fossil whales.

In a study published in 2021 by Italian researchers (Collaretta et al. 2021) including also Giovanni Bianucci, who co-authored the 2007 review paper, it is demonstrated that research on fossil whales goes much further back in time. And involves no one less than the famous Italian artist, scientist, inventor and basically all-round genius Leonardo da Vinci. They provide very convincing arguments that da Vinci personally observed and studied a fossil whale skeleton in Tuscany in the late fifteenth century (around 1480), making first-hand observations and clearly recognizing it as the skeletal remains of an ancient and probably marine animal.

So all of a sudden the start date for the study of fossil Cetaceans is postponed by almost two centuries. And we are made to believe that no one ever found a fossil whale bone, pondered on it and maybe even published about it between the account of da Vinci in 1480 and the work of Scilla in 1670? That the entire civilized world was populated by ignorant dumbasses, who just did not care or were literally too stupid to recognize a fossil? Because fossil whales aren't that rare, particularly in the Tertiary sediments of Europe. They belong to the most common and easily recognizable fossil mammals found.

So the question, of course, must be asked: is the story we are provided with here true, and there were two centuries of complete and utter ignorance? And this exactly during the time often aptly named as the 'Renaissance', when supposedly people with enough leisure, money and incentive were starting to dig up antiquities of all kinds all over the place and put them in their cabinets and castles?

Or are we just missing the largest part of the picture because works from the time in–between have been completely lost, suppressed or have just not been studied yet, being hidden in the basements of some libraries and never looked at by modern researchers?

One may argue that this is highly unlikely. Yet da Vinci's description of his encounter with the fossil whale apparently escaped the notice of all researchers up to 2014, when it was first suggested that it was exactly this that he saw (Etheridge 2014,

2019). Now, da Vinci certainly isn't an 'understudied' author. And if even in his surviving texts (and it is very likely that not all he ever wrote has been preserved to the present day) there is still new stuff to be found that fundamentally affects our understanding of the history of vertebrate palaeontology, how much by lesser known or totally forgotte n authors of the past may still be out there?

This is what I call 'the deep time of science'. Just like as a palaeontologist you have to dig through the sedimentary strata to find the fossils and reconstruct the probable history of life on earth from these fragmentary remains, as a historian of science you must dig through the 'cultural layers' and work with the fragments that have been preserved of the research endeavours of previous generations to reconstruct the development of thoughts and understanding.

CHAPTER 2

American oddities

O f particular interest to the central thesis advocated here, namely that modern vertebrate palaeontology was the result of a 'Great Reset' of civilization that took place during the late eighteenth and the early half of the nineteenth century, are the United States of America. The USA was the first country in the world to completely do away with the traditional monarchic and aristocratic, and to some extent clerical, power-elites and to become, in many ways, a 'modern country'.

The founding fathers, almost exclusively Freemasons, put up a large societal experiment in the new world, which over the course of the following almost two and a half centuries became an increasingly important power centre, until it emerged as arguably the last remaining military and economic juggernaut in the world of today, in particular dominating the entire planet with its empty shell of a 'culture', consisting mainly of Hollywood and Disney movies, fast–food restaurants and Pop–music. A dominance of very questionable value, that, unquestionably, no other country in the history of the world has ever enjoyed.

The current territory of the USA is an amalgamation of former European colonies and land previously inhabited only by 'natives', for a long time portrayed as particularly primitive and barbarian people who had to give way for the establishment of the new 'enlightened' order brought to them in very often not too enlightened means by the US government. The spreading of the benefits of 'freedom and democracy' were first felt by these people, long before the USA started to bomb random countries in Africa and the Middle East for allegedly the same reasons.

The entire success story of the USA in the late eighteenth and nineteenth century is built – rather than on the 'pioneering spirit' and all kumba-yah 'pursuit of happiness' – on four fundamental pillars:

1. The destruction of the old colonial power structures and replacement of the traditional aristocratic power-elite by a new one made up largely of 'pioneers' and 'entrepreneurs' wo evolved into the capitalist 'robber-barons' and banking cartels of the industrial revolution times, starting with the War of Independence and continuing up to the American–Spanish war of 1898 (and beyond).

2. Expanding its territory and sphere of influence by constant aggression towards both the native population and other independent countries. Examples of the latter include the American-Mexican war of 1848, pressuring Japan into opening towards the West by Perry's expedition in 1853, the annexation of the Kingdom of Hawaii in 1898, the American-Philippine war of 1899–1902 and too many others to be mentioned here.

3. Genocide and deportation of the native population.

4. Slavery.

After World War 2, to this can be added:

5. The complete intellectual rape and outright theft of all of German technology and scientific advancements on a scale

never before seen in the history of the civilized world, and the 'repatriation' of German elite scientists, technicians and intelligence and military experts to the US after 1945 during 'Operation Paperclip' and many other, less well known and more secret projects. This formed the basis not only for the US space program, but also helped them establish the most efficient intelligence agencies in the world and made a major impact on all kinds of military and technological development, which has ever since made the USA the leading power in these areas. Very little of it was genuinely 'made in the USA', though. A very major part was in fact 'made in Germany'.

So this is the legacy of the country that likes to pose as the 'beacon of enlightenment' and the 'peak of western civilization', the one which became the first to establish the 'New World Order', so proudly announced, complete with masonic symbolism, on every single note of the American dollar, which since then has spread over the majority of the globe. A legacy of slavery and genocide, never-ending wars of aggression and theft, built on the death and misery of uncounted millions, all under the guise of 'freedom and democracy', or 'liberté, égalité, fraternité', the ideals of the enlightenment, the buzzwords of the French Revolution, and the ones with which all kinds of heinous crimes against humanity are cloaked for the stupid public on a daily basis up to the present day.

America in the late nineteenth and early twentieth century was a bunch of former European superpower colonies that themselves became a superpower with a worldwide colonial empire, and 'European imperialism' had de facto been fully embraced by the American government and public.

The American Empire stretched all over the globe and included many overseas possessions that were certainly not 'American' at all. In 1941, when the USA entered the Second World War against those evil jingoist Japs and Nazis, it was itself a major imperial power. The 'greater United States' comprised Alaska and Hawaii, bought from Russia in 1867 and annexed in 1898 respectively (both achieved statehood as late as 1959) as well as Cuba, Puerto Rico, the Philippines, former Spanish colonies annexed by the USA after the 1898 American–Spanish war.

In the Philippines, a gruesome war ensued against the rebelling native population. Now mostly memory–holed it lasted from 1899 to 1902, and nobody ever counted the civilian casualties, but estimates by historians go up to one million dead, which would have been an eighth of the entire population of the Islands at the time.

But America's overseas possessions did not end there. Additionally, many Islands in the Pacific and the Caribbean were (and some still are) simply American colonies, the Northern Marianas, Guam, the American Virgin Islands. The Panama Canal Zone was American territory, and it can be argued with good reason that the entire country of Panama was nothing but an American protectorate, as was Liberia in Africa.

Add to this the territories snatched from Mexico after 1848 – the entire states of California, Utah, Nevada, Arizona, as well as parts of Colorado, New Mexico and Wyoming (yes, Mexico extended as far up as Wyoming before the war of 1848) and the picture that emerges is that of an aggressive land–grabbing colonial superpower easily comparable to and in fact surpassing most of the European Empires of the day. Ruling over uncounted millions of non–Americans and people of colour all over the world, just like the British or French or German Empire did before the World Wars. It is still pretty much the same today, although American influence is less ensured by direct rule than by the omnipresence of the American military, with its bases spread all over the world.

These facts had to be stated clearly here at the start of this chapter, because the role the USA played in the major civilization reset can by no means be underestimated. In many ways, it served as the model for the rest of the world, having a head start and establishing the new power structures that we see today all over the place. And the simple fact that the majority of the rest of the world has by now adopted this model (more or less peacefully) just underlines how central and important the USA was and still is.

In the eighteenth century, before the 'War of Independence' of 1778, things were completely different. The North American continent was under colonial rule by European monarchies, Great Britain, France and particularly Spain. The territorial shifts that took place between the various powers in the eighteenth and early nineteenth century are enormous. The United States, despite the Louisiana purchase of 1803, remained a minor player – at least considering their territory – up to right after the end of the Napoleonic wars and their aftermath in Europe.

The largest American territories in 1778 were held by the great worldwide Empire of the Kingdom of Spain, itself only a remnant of the once much vaster Holy Roman Empire of the German Nation under Emperor Karl V. upon which the sun never set. The Spanish territories stretched northwards to nowadays Alaska and eastwards to the Mississippi and beyond, even including Florida and the coast of Texas.

France under Napoleon got hold (again, one must say, because these territories had already been French a long time ago before the Seven Year's War) of some of the Spanish territories (more than 2 million square kilometres west of the Mississippi) in 1800 in exchange for the Duchy of Parma in Italy which went to Spain in return.

Napoleon sold them almost immediately to the United States in 1803, allegedly to finance his ongoing and seemingly endless all–out war against Europe. The sum of 15 million dollars (which would be about 250 million dollars today) is absurdly low for what was effectively the largest purchase of land in recorded history, by which the USA suddenly doubled its territory, and the entire story is more than fishy. But the USA also bought Alaska in 1867 for only 7.2 million dollars (about 150 million dollars today), so it seems they were constantly 'lucky' in expanding their territory by excellent bargains.

Be this as it may, the major point in all this is that the century-long colonial rule of European superpowers of the day should, in some way, have resulted in the discovery of fossils from the territory of the United States. Large parts of the country are so stacked with fossils, most spectacularly dinosaurs and Tertiary and Quaternary mammals, that it is impossible to miss them. How many of them were unearthed in almost zero time by just a handful of scientists – it basically comes down to Leidy, Cope and Marsh – during the 'bone wars' speaks for itself.

These fossil vertebrates even figured prominently in some of the myths of several native people (see Mayor 2007 for a comprehensive overview). Alas, if you look at the official literature, there is nothing. Nothing at all. Not even a mention in passing of some odd shaped rocks resembling the bones of giants or anything. It appears that the French, Spanish and British were just completely and utterly ignorant, or did not care, or were too dumb to recognize a huge bone or skull sticking out of the rock somewhere.

George Gaylord Simpson, one of the founding–fathers of the modern neo-Darwinian paradigm of evolution and one of the leading palaeomammalogists (researcher on fossil mammals) around the mid of the twentieth century, gave a very detailed account about the early days of American vertebrate palaeontology in 1942 (Simpson 1942).

He relies heavily on the invaluable Bibliographies of American vertebrate palaeontology that were compiled painstakingly by

Oliver Perry Hay in 1902 and (as a much extended second edition) in 1929/30. Hay's work is the holy grail of any historical research on the early days of American vertebrate palaeontology, compiling everything from the first publications ever throughout the classic efforts of Harlan and his contemporaries, the age of Leidy, Cope and Marsh (the 'bone wars' years) to the emergence of the modern face of American palaeontology in the early twentieth century, which has not changed much ever since.

And what story does Simpson's (1942) account tell us? To put it short, it is that basically nothing happened before the founding of the United States of America, and that even after that it took until the end of the Napoleonic wars (and their aftermath) until things got really started. The wars starting with the French Revolution and culminating in Napoleons campaigns lasted from 1789 to 1815 in Europe. That is more than 25 years of basically nothing but constant, ever expanding war (or being, if not at war, a defeated country under French military rule which wasn't all that benevolent).

These immense gore fests, never seen since the 30-year war of 1618–1648 in Europe and only surpassed in brutality and loss of life by the two twentieth century world wars, were mirrored in North America by the Mexican war of independence, lasting from 1810 to 1821, and the second American–British war of 1812–1815. In South America, there was also conflict all over the place in the Spanish colonies. Napoleon had started his Peninsular War campaign against the kingdoms of Spain and Portugal – at that time still two of the leading European colonial powers, which held control over most of the Americas – in 1807. First, Spain and France together attacked Portugal in 1807, but Spain found itself stabbed in the back by its former ally Napoleon in 1808 and soon was entirely occupied by French forces.

As a reaction to the breakdown of the Spanish kingdom in Europe, a consecutive series of wars that lasted from 1808 to 1833 broke out all over the American colonies. The only exception was Brazil, which was under Portuguese rule, and harboured the Portuguese royal family in exile for 13 years from the Napoleonic invasion of 1807 onwards. Brazil became fully independent in 1825, but instead of fighting the monarchy, the Brazilians fought to keep it, and with Prince Pedro of Portugal becoming the first Emperor of Brazil, relations to the former colonial overlords remained comparatively amicable for an extended time span afterwards.

Probably surprising to the champions of 'liberal western democracy' of the US and French brand, the Imperial rule made Brazil an up–and coming great power in the nineteenth century, being surprisingly modern and advanced, particularly in comparison to any of the former Spanish colonies, who had largely become 'democracies' following the 'New Order' as promoted by the USA and revolutionary France, and were, frankly speaking, a number of backwards shitholes compared to Brazil.

Imperial rule in Brazil ended in 1889, and thus ended the short episode of Brazil as an emerging world power that was the only American country ever to have a chance to become a serious rival to the USA. Nowadays, Brazil is just as much of a borderline Third World country as the rest of Latin America, but enjoys the benefits of American style 'freedom and democracy' in return.

Disregarding Brazil, which just had a brief and not-too-bloody war of independence in 1824, the rest of the Americas was entangled in a decade-long world war to varying extents just as much as the entire continent of Europe was, and it emerged as something completely different from what it was before. The USA had consolidated itself as the major power in North America, all the former Spanish mainland colonies had become independent, Brazil was an independent Empire with close ties to Portugal.

It is easily understandable that during such troubled times, there is very little incentive for anyone to do scientific investigations, unless they are tied to matters of military or economic importance, and it is therefore also unsurprising that it took some time for vertebrate palaeontology to get back on its feet – or even really start – in the Americas.

It remains mysterious and inexplicable to me, however, that in the entire time from the late 15th to the late eighteenth century, there was nothing going on in all the Americas, even though not only the USA, but also many South American countries, particularly Argentina and Brazil, are filled to the brim with some of the most spectacular fossils in the world which should have attracted the interest of scientifically minded people, explorers or just curious ordinary Joe's just as much as the comparatively much more meagre findings in Europe did. But nothing officially happened until very late in the eighteenth century.

Simpson's (1942, p. 130) account deals with all the discoveries prior to the late eighteenth century in a couple of sentences: "The first vertebrate fossils to be seen by Europeans in the Western Hemisphere were mastodon bones collected by the Indians in Tlascala, and shown to Cortez's army in 1519. A few casual finds were made in the next two centuries, but these also had no sequel and cannot be called scientific discoveries. The find that may be considered the true discovery in this historical sense was made by a party under the Canadian Charles Le Moyne, second Baron de Longueuil, on the Ohio River in 1739. Mastodon bones were taken to France by Longueuil and studied by Guettard, Daubenton, Buffon, and others. Another important collection was made by Croghan in 1766 and sent to London. It was not until about the end of the eighteenth century that it was firmly established that most of these large bones represented a distinct, extinct, herbivorous species allied to the elephant – the species now called *Mammut americanum.*"

Simpson (1942, p. 134) makes a further, very interesting statement. Coming back to the supposed Mastodon bones shown to Cortez by the Indians in 1519 in the vicinity of Tlaxcala, a region of Mexico rich in fossil mammals, he notes: "It is unlikely that such striking objects escaped the attention of the early natural historians of the Spanish possessions, who did give notices of fossil bones in South America, at least. A search of that literature may produce interesting data also on Mexican discoveries, but limitations of space and time have prevented me from undertaking so considerable a task regarding a topic that is only incidental to my main theme." So basically neither Hay nor Simpson (nor anyone else ever since) appears to have undertaken the task to check for such accounts, and so it is unsurprising that none of them are known.

Is (or was) there an entire body of early palaeontological research lost to us? Spanish scientists working in the North American West and South America for centuries, their reports and accounts destroyed, forgotten or hidden away in the dusty basements of some ancient Spanish or Latin American archives?

The first ever illustration of a vertebrate fossil from the Americas was published – or officially so – in 1756, when the French naturalist Jean–Étienne Guettard (1715–1786) figured a molar tooth of an American Mastodon (*Mammut americanum*) (Guettard 1756), being at a loss to determine what kind of animal it could have belonged to. His French colleague Louis Jean–Marie Daubenton (1716–1799) re–examined the dental and skeletal material and made comparisons to the findings of the Siberian mammoth (*Mammuthus primigenius*) as well as to extant elephants, particularly stressing the similarities of the thigh-bones (femora) (Daubenton 1764).

15

So here we see Cuvier's method of comparative anatomy cleverly and excellently employed long before Cuvier was even born. Simpson (1942) also has to admit, that Daubenton's study effectively employs the 'Cuvierian' method, and that Daubenton, if anyone, should be credited with it: "This is one of the four most basic discoveries or principles in the rise of vertebrate palaeontology, and it may fairly be dated from Daubenton, although even he had less important predecessors (such as Catesby's slaves) and although it is usually credited to Cuvier a generation later." (Simpson 1942: p. 145)

Simpson adds a footnote, in which he says the following, which I quote in full because of its outstanding importance. Remember that Georges Gaylord Simpson was one of the godfathers of both vertebrate palaeontology and evolutionary theory in the twentieth century, and that him coming basically to the same conclusion as promoted in this book is certainly noteworthy: "He [Cuvier] may justly be honored as the father of vertebrate palaeontology, but it is improbable that Cuvier originated any of the truly fundamental principles of that science. He developed certain of these principles, exemplified them, systematized them, and for the first time brought together a considerable mass of material data. These great achievements warrant and assure his pre–eminent place in scientific history, without claiming for him priority on particular points that were definitely anticipated." (Simpson 1942: p. 145).

So what Simpson says here, plain and simple, is that Cuvier did not invent anything new in vertebrate palaeontology, at least not anything of fundamental importance. That instead he was mainly a compiler of data brought together (and often, as we will see, already published) by previous authors, and that his main achievement was to make these data more accessible to an emerging 'intelligentsia', particularly through the many editions of his *"Recherches sur les ossements fossiles"*.

Cuvier became, as I said above, the most prominent figure, the 'superstar' or 'poster–boy' of the 'new science'. This 'new science' could easily be sold to the public as a triumph of 'enlightenment' and the ideals of the French (and by extension American) revolutions, while in fact everything was already there long before. The new emerging power elites of revolutionary and post–revolutionary Europe and America just appropriated the centuries of research done before, under the 'Ancien Régime', and Cuvier (and some of his contemporaries, none of whom, however, ever reached his prominence or influence) were the central figures in this appropriation.

But coming back to Simpson's footnote, what is the story behind 'Catesby's slaves'? Here we get into almost uncharted territory, that further puts serious questions to the prevailing narrative. Mark Catesby in 1743 published the second volume of "*The natural history of Carolina, Florida and the Bahama Islands.*" There (Catesby 1734–1747, Appendix, p. vii) we read the following account: "All parts of Virginia, at the Distance of Sixty Miles, or more, ["from the sea," evidently omitted, *remark by Simpson*] abound in Fossil Shells of various Kinds, which in Stratums lie imbedded a great Depth in the Earth, in the Banks of Rivers and other Places, among which are frequently found the Vertibras, and other Bones of Sea Animals. At a place in Carolina called Stono, was dug out of the Earth three or four Teeth of a large Animal, which, by the concurring Opinion of all the Negroes, native Africans, that saw them, were the Grinders of an Elephant, and in my Opinion they could be no other; I having seen some of the like that are brought from Africa."

So it is not Daubenton who has to be given credit for first employing the method of comparative anatomy in vertebrate palaeontology, but Catesby and a number of never to be named, never to be known native Africans, brought to America as slaves to waste away for their overlords, who were perfectly able to recognize these fossils as what they were, from their deep knowledge of the animals and nature of their African homelands, remains of elephants that once roamed the great North American continent where no living man had ever seen an elephant.

This story is highly significant in various ways. Firstly, it completely destroys Cuvier's claim as being the 'father of comparative anatomy in palaeontology', even more so than the study of Daubenton (who could not lay much claim to having invented the method, because he died shortly after Cuvier started to put out his first publications. Ironically, on the 31st of December 1799, right at the end of the eighteenth century).

It also puts another claim of Simpson (1942, p. 132) into perspective, where he refers to the (well documented) fossil collecting activities of various Native American tribes: "Pre–Columbian Indians certainly found and occasionally collected fossil bones, but in general these discoveries are no real part of palaeontological history. Various reported Indian legends of fabulous beasts represented by fossil bones have little ethnological and no palaeontological value; the data are sparse, often untrustworthy, and carry little conviction of genuine and spontaneous (truly aboriginal) reference to real finds of fossils claim."

So, on one hand we have Catesby's 'Negro slaves', who are perfectly fit to confidently identify an elephant tooth when they see one, even if it is as distant from a living elephant as an American Mastodon, and, if the story is true, could lay claim as to be the first people to ever employ the method of comparative anatomy in vertebrate palaeontology.

On the other hand, we have the 'Pre–Columbian Indians' who, allegedly, were not able to do so and basically are not supposed to have had any idea about what they had just found or collected.

These people, roaming the entire American continents for uncounted millennia, must have found fossils almost on a daily basis in many areas, like the American Midwest, where they are sticking out of practically every rock.

Particularly the native people of South and Middle America are credited by conventional archaeology and history for developing some of the most advanced cultures of their time, with considerable astronomical and mathematical expertise, and they certainly also had an excellent knowledge about the plants and animals in their domain.

The architectural relics of these people belong to the greatest monuments found anywhere on earth and have amazed countless generations by their sheer grandeur, beauty and craftsmanship. Their calendars were, so archaeologists tell us, more exact than the ones in use at the same time in Europe. Yet these people were allegedly too stupid to identify a bone dug out of the ground? Or maybe able to recognize that these bones are different from the bones of all the animals with which they were familiar with, which they domesticated, or hunted, and butchered for food, so being completely aware of their anatomy, including the skeleton? Obviously. Conventional history also tells us that they were too dumb to invent a wheel, although they built the wonders of Macchu Picchu, Tiahuanaco, Tenochtitlan, Sacsayhuamán and Puma Punku.

This is the story with which we are presented over and over again. Palaeontology as we know it was invented in the early nineteenth century, and Baron Cuvier was the central figure in the establishment of the 'new science' during and after the great revolutions of the late eighteenth and early nineteenth century.

Everything before was just fumbling around with no clue, basically at the level of Johann Jakob Scheuchzer (1672–1733), the Swiss physician who identified the skeleton of a giant Tertiary salamander (*Andrias scheuchzeri*) as that of a poor sinner who had drowned in the deluge (Scheuchzer 1726), and of course it was the genius of Cuvier who correctly identified the remains for the first time.

And outside of Europe, there was just nothing going on. Primitive people they were, regardless what their other officially admitted achievements may have been (which can hardly be denied, as we are facing the archaeological relics and the grandiose architecture of these ancient civilizations, which just could not be destroyed). So yes, they built great structures, wonders of the world, but apart from that they had no idea of the world in which they were living and no ability to do 'science' as the new order, established since the nineteenth century, defines it.

They were haplessly stumbling around, literally surrounded by fossils, Tertiary mammals and even dinosaurs practically biting them in the ass on a daily basis, but they never bothered to investigate, never gave a second thought. They were like children, waiting for a new breed of white men to bring them the benefits of civilization.

And not only this, but, as we have already seen, even the Spanish Colonial Empire was totally incapable of achieving anything at all. They were just sitting in most of the Americas for centuries, twiddling their thumbs. It needed Anglo-Saxon protestant explorers with a new democratic and enlightened attitude to reveal the treasures of the North (and South) American earth. Nobody else was ever able to do it.

Just how believable is this scenario? Just how believable is it with respect to the case of 'Catesby's slaves'? Were ancient civilizations in the Americas, Africa and Asia, as well as century-old European superpowers totally incompetent dumbasses with an intellect below that of a bright kindergartener? What amount of IQ does it take to recognize a bone as a bone, a seashell as a seashell?

Apparently it was an extremely great feat even for white Anglo–Saxon Europeans to reach that point, and we only fully managed to do so a mere two hundred years ago, because before that we were basically still living in a dark age, dominated by superstition (let's call it the 'Scheuchzer age'). And it was completely impossible for any non-white civilization on Earth. They may have built the most astounding pyramids, temples, pagodas and mosques, but preoccupied with creating some of the greatest architecture on Earth, these dumb fucks were never able to recognize a bone or a seashell.

What, on the other hand, are the chances that a lot of ancient knowledge is just lost in time? Probably not even surviving in written form, because it all was deliberately or accidentally destroyed or has long been forgotten? How many manuscripts and documents of the Spanish colonial era may have disappeared during the ceaseless wars from 1808 to 1833? And the many more endless wars in Middle and South America ever since, some of them as brutal as the War of the Triple Alliance (1864 to 1870) which almost genocided the entire country of Paraguay? What was probably destroyed or stolen from Spain during the Napoleonic occupation from 1808 to 1814?

The French weren't particularly reluctant when it came to collecting spoils of war, we will see that in the case of the 'Grande animale de Maestricht', a story which also is a smoking gun that indicates just how important vertebrate palaeontology was in the context of what was going on in the world at that time. That the establishment of the 'new science' was of great importance, and that, like a spider in his web, again it is Monsieur Cuvier whom we meet in this story in a very prominent role.

We will investigate these and many other things in the following chapters, and we will get back to America on various occasions. Just keep in mind some of the oddities and open questions we have already encountered. There are many, many more to come.

CHAPTER 3

*The myth of the ignorant
Spaniards – early vertebrate
palaeontological work in Argentina*

A rgentina is one of the countries in the world graced with some of the most astounding fossil vertebrate sites. Numerous species of dinosaurs and other fossil reptiles have been discovered there, and the number of excellently preserved fossil mammals, many of which, like the giant armadillo Glyptodon and the elephant-sized ground-sloth Megatherium, have become icons of palaeontology known even to non-specialists, is almost unlimited.

It is therefore fascinating to look at the first chapters of vertebrate palaeontology research going on in the country, as far as it is still reliably documented. Argentina was one of the first (and few) non-European countries to establish a palaeontological tradition of its own during the nineteenth century. This was mainly the result of the work of both expeditions from European researchers (not to the least a certain Charles Darwin), and scientists who immigrated to Argentina from Europe.

One of the most prominent of these was the German naturalist

Hermann Burmeister. Born in Stralsund, on the coast of the Baltic Sea in 1807, Burmeister was an old school polyhistor, whose scientific endeavours started out mainly in the field of entomology, but soon he also became a very capable palaeontologist, particularly studying fossil vertebrates. He was a personal acquaintance of the great German explorer Alexander von Humboldt, and Humboldt's legendary travels and discoveries in South America may have been a major influence to prompt him to go to Argentina in 1861.

When he arrived there, the entire country was in civil war, and only in 1862 he could attain the position he was promised, directorship of the Museo Publico in Buenos Aires. There he worked ceaselessly until his death in 1892. When he died, he was some kind of national hero. The Argentinian president himself was leading the procession of the state funeral with which he was honored.

Burmeister worked tirelessly and unearthed numerous remains of extinct creatures from the Tertiary and Quaternary Formations of the huge country, while also working in many other scientific fields. It is almost unbelievable what he achieved in the short time he was active there, but he paved the way for later generations of native Argentinian palaeontologists, most prominently his pupil Florentino Ameghino (1853–1911), who produced one of the largest bodies of work ever by a single vertebrate palaeontologist and became the true godfather of South American vertebrate palaeontology.

But vertebrate palaeontology in Argentina started long before Burmeister and Ameghino, and it goes back far into the Spanish colonial times in the eighteenth century. The single example that is halfway known in the scientific community is the first ever nearly complete skeleton of the giant ground sloth *Megatherium americanum*, which became widely known by the end of the eighteenth century in European intellectual circles as 'The great animal from Paraguay' (although it was found in Argentina). On exhibit in the Museum in Madrid, it became the first fossil vertebrate skeleton ever completely mounted in the world. The story of the Madrid *Megatherium* is such an important one, central to so many topics of this book, that it will be dealt with in the next chapter.

A very interesting article on the early history of Argentinian palaeontology under Spanish colonial rule was published in 1992 (Schávelzon & Arenas 1992). Of course, it discusses the Madrid *Megatherium* at some length, but it also provides more glimpses into the earliest (currently documented) endeavours that are not easily found anywhere else. The famous *Megatherium*, found and excavated at Lújan in 1787 and sent to the King of Spain in 1789 was certainly a peak of Spanish colonial vertebrate palaeontology in Argentina, but it was far from being the start.

The article refers to very early records of 'giant bones' found in South America, dating back to the early sixteenth century. Interestingly, even the famous Half–Inca Historian Garcilaso de la Vega, son of a niece of the Inca ruler Huayna Capác and a Spanish conquistador, mentioned these bones in his own account of the history of Peru, "*Comentarios Reales de los Incas*", published in Lisbon in 1609. It is therefore evident that both the Spanish colonial rulers and the native people were fully aware of the presence of these fossils long before the eighteenth century. But apart from a few mentions in the surviving literature, very little of the knowledge that they probably had about these things has survived to the present day.

In 1766, more than twenty years before the famous discovery of the 'Great animal from Paraguay', the first *Megatherium* skeleton, a considerable amount of fossil bones was collected and deposited in Buenos Aires. A Spanish frigate captain, Estevan de Alvarez Fierro, had recovered numerous bones of gigantic ancient animals, forty leagues from the town of Arrecife (today called Arrecifes, in the province of Buenos Aires).

It is indicated that the excavation work was accompanied by detailed drawings of the finds in situ. This was also done during the excavation of the *Megatherium* skeleton. The bones were apparently carefully packed in leather bags and thus safely transported to Buenos Aires. This account is very interesting for the fact that the Spaniards obviously employed rather sophisticated and 'modern' excavation methods, painstakingly documenting and securing the finds, already at such an early date, something that was very rarely done even more than a century later during the American 'bone wars'.

Three local experts, Matías Grimau, Ángel Castelli and Juan Parán, all of them surgeons and doctors, were appointed to study the finds and testified that they were genuine remains of ancient animals, or maybe 'giants' Sadly, as Schávelzon & Arenas 1992 state: "From here we have no more information about it, and it is most likely that the bones never reached Spain, since otherwise we would have the documentation here or there. This file was forgotten for many years, until it was published by Miguel Navarro Viola in his Revista de Buenos Aires together with some notes by José María Gutiérrez in 1866. Much later it was remembered again by Alberto Palcos (1944) and more recently it was transcribed in part by Julián Cáceres Freyre (1968)."

So here we have undoubted evidence of palaeontological research on a very sophisticated level conducted by the Spanish colonial power in South America, which has been almost completely memory–holed and of which only a chance documentation remains. What exactly the fossils were and if anything else came of their study, we will probably never know. But it is obvious that it was treated as a matter of great importance and that a considerable amount of time and work was devoted to it.

How many similar cases may have occurred during the long time of Spanish colonial rule in the Americas, about which we know exactly nothing? How many fossils may have become lost to science, how much knowledge may have been there already centuries ago that is now completely gone? It is by sheer luck that we have at least this single account, but it is out of the question that this could not have been a singular event.

The next big step, however, was allegedly not made by a native inhabitant or one of the Spanish colonial rulers, but by a travelling Englishman, Thomas Falkner (1707–1784). Falkner was a Jesuit, working as a missionary, but he was also an accomplished physician. He travelled in Patagonia for almost forty years of his life, and published an extensive account of his travels in 1774 ("*A description of Patagonia and the adjoining parts of South America.*").

At the banks of the Carcarañá River, near the city of Santa Fé, he encountered numerous fossil bones. He tells, among other finds, about a fossil that unmistakeably could only be the shield of a giant armadillo of the genus *Glyptodon* or a closely related form: "I myself found the shell of an animal, composed of little hexagonal bones, each bone an inch in diameter at least; and the shell was nearly three yards over. It seemed in all respects, except it's size, to be the upper part of the shell of the armadillo; which, in these times, is not above a span in breadth." (Falkner 1774, p. 54).

This definitely is the first – known – account of a find of these animals in Argentina, which were absolutely ubiquitous and whose fossils belong to the most common fossil mammals encountered all over South and even Central and parts of North America. Falkner goes on to describe other finds he and others have made, including the complete skeleton of a 'monstrous alligator'. Were any of these fossils excavated? Were they ever deposited in a museum? Were they ever studied? We have no idea, the only thing remaining is Falkner's account. Have they all been lost forever during the cataclysm of the South American 'War of Independence' and its aftermath, a never ending series of conflicts, revolutions, civil wars and all–out wars ravaging through large parts of Central and South America for most of the nineteenth century? Again, as with the findings of Fierro in 1766, we are left with nothing, except that Falkner's detailed description allows us to identify one of his discoveries with confidence as that of a glyptodontid.

Of course, Wikipedia (as of the 15th of June 2022) tells us that: "the first mention of the genus *Glyptodon* in Europe was in 1823, from the first edition of Cuvier's *Ossemens Fossiles*". Once again, it is, of course, Cuvier who is credited with the discovery, even though he only had information on a single femur and part of a tail at his disposal which he mistook for *Megatherium*. Interestingly, he never saw the original fossils, but had to rely on a letter provided to him from Dámaso Antonio Larrañaga (1771–1848), a local fossil collector (among many other things, like being a priest, an architect and a renowned botanist, who happened to be largely responsible for funding the National Library and the National University of Uruguay, all at the same time, again one of those notorious almost impossible nineteenth century polyhistors) in Montevideo. Nor did Cuvier ever see the original skeleton of *Megatherium* in Madrid.

In fact, as seen above, the discovery of the glyptodons dates back almost half a century before Cuvier, and must be credited to Thomas Falkner, who even correctly identified the remains as being those of a giant armadillo, whereas Cuvier blundered and had no real idea what he was dealing with, which, to his credit, is somewhat understandable with such scarce remains. Also, the debate whether *Glyptodon* and *Megatherium* were two different animals, whether *Megatherium* was more closely related to sloths or armadillos, and whether it was heavily armoured or not continued well into the late 1830s and was just finally resolved by Richard Owen in 1839.

However, Cuvier being known at his time as 'the man who could restore an entire animal from just a single bone', certainly not showed the most masterful display of his abilities, even though he was eventually proven right in his idea that *Megatherium* was more closely related to sloths than to armadillos as many other experts had claimed. It was indeed Cuvier's fellow Frenchman, Alcide d'Orbigny (1842) who was the first to point out that the description of Falkner could refer to nothing else but a *Glyptodon*, but this was long completely forgotten or memory-holed by the official history of vertebrate palaeontology.

It is also very unlikely that Falkner was the first one to see fossils of a *Glyptodon*. They are so commonly found in the Pampas Formation of Argentina, their VW–beetle sized armours and skeletons strewn all over the place, that one has to be completely blind not to recognize them. And they are so similar to extant armadillos, particularly in the construction of their carapace-like shields, animals that are still widely found in South America, that it also would be close to impossible for either a Native American or a Spaniard not to immediately see the similarity and make the only viable conclusion, the one which Falkner also reached, namely that these things were nothing but an ancient form of giant armadillo.

However, apart from a few fragments, it officially took until the voyage of the H. M. S. Beagle, when a certain Charles Darwin collected numerous fossil remains of extinct South American land mammals in both Argentina and Uruguay (Darwin 1838), including more of *Glyptodon* (but largely a closely related form today known as *Neosclerocalyptus*), that a realistic picture of the animal emerged.

Richard Owen (1839) described most of what was known about *Glyptodon* at the time – not yet including Darwin's finds which he described later – in the first detailed monograph of the animal, but he had no complete skulls or skeletons at his disposal. This all happened at a time before the *"Origin of species"*, when young Owen and young Darwin were still on friendly terms. Owen's 1839 paper makes no mention of Falkner's discovery (but discusses Cuvier's attempts from 1823 in detail), although Falkner's work was well known at the time (and it is certain that at least Charles Darwin had read it, as he mentions Falkner explicitly in his account of his voyage on the H. M. S. Beagle).

Darwin's discoveries were indeed remarkable. His friend Owen, who described his fossil collection, was able to identify no less than eight new species of fossil mammals from the remains that Darwin had unearthed. Not to mention the fact that these fossils may have had a major impact on the creation of the theory of evolution by natural selection, so that the comparatively small and not too well-preserved collection of fossils that Darwin accumulated was even bestowed the honour of being the topic of an entire book recently (Lister 2018).

So a single expedition and the work of a single Englishman easily had outshone centuries of Spanish colonial rule, which only resulted in the discovery of a single species, *Megatherium americanum*, which wasn't even named by a Spanish scientist but by the inevitable Georges Cuvier (1796a), in one of his first ever publications.

Even more remarkable is the fact, that after independence from the Spanish crown, Argentina became the target of a number of explorers and dealers, who exported large quantities of fossil bones from the country and sold them to European scientists and research institutions (Podgorny 2012).

It is obvious that even before Darwin's discoveries, it was well known that the country was full of fossil treasures, and now, in the 'new world' of the post–revolutionary era, they could be sold for good money to researchers willing to pay. This is particularly interesting with regard to the fact that, officially, so little was known about the fossils of the country at this time. And of course, it shows how fossils had suddenly become just another 'ware' that could be traded, just like anything else.

Commerce had now a say in what was once the assembling of national treasures, the pride of kings and princely cabinets, like the original *Megatherium* skeleton in Madrid, and it is sad to see that even Alexander von Humboldt's faithful travel companion, Aimé Bonpland, had become deeply engaged in these activities.

So this is what little we are officially supposed to know about vertebrate palaeontology in Argentina before the 'Great animal of Paraguay', the prize specimen of Spanish colonial palaeontology, was unearthed in 1787. This very interesting and highly enlightening case will be dealt with in the next chapter.

CHAPTER 4

*The great animal of Paraguay
which came from Argentina, enter
Georges Cuvier, plagiarist*

A major smoking gun concerning vertebrate palaeontology in the Spanish colonies of the Americas, and basically the history of the entire discipline during the late eighteenth and early nineteenth century, and how narratives were changed and pushed into certain directions, is the case of the type skeleton of Megatherium americanum, still at display today in the Museo Nacional de Ciencas Naturales in Madrid.

The story, to abbreviate it as much as possible, because it would well be worth an entire book of its own, goes as follows. The skeleton was discovered in 1787 by a Dominican friar, Manuel de Torres. He asked the viceroy of Rió de La Plata, the Marquis de Loreto, to send out a draughtsman from the Spanish Royal Artillery Corps to draw and document the skeleton. See how well that ties in with the story of the 1766 discoveries by Captain Fierro, just as if this kind of documentation of such important fossil finds was a long–established, common practice in the Spanish colonies?

The skeleton was accordingly drawn and documented extensively before it was excavated. According to Podgorny (2012) the skeleton, after it had been transported to Buenos Aires, was originally already mounted there, before it was, again, disassembled, shipped to Madrid and remounted in the Spanish capital.

This, in itself, is absolutely astounding. The type skeleton of *Megatherium* is officially acknowledged to be the first ever skeleton of a large fossil vertebrate to have been 3D–mounted in any museum in the world. That this should have been done two times, once in Buenos Aires, and a second time in Madrid, in the late 1700s is just mind–blowing. The first ever mounted dinosaur skeleton was Leidy's *Hadrosaurus foulkii* in 1868, more than half a century later. And the mount, which was assembled with the help of famous artist Waterhouse Hawkins and a certain Edward Drinker Cope, not even included a single original bone. It was all a combination of plaster casts and guesswork, while the Madrid *Megatherium* was the real thing.

The mounting of *Megatherium* becomes even more of a mystery if one takes into account why it was declared by the Spaniards that it was impossible to make a cast of the skeleton (Piñero 1988) in 1833, when the British ambassador in Madrid requested it. As Piñero (1988, p. 163) states: "The Society for the Preservation of the Royal Museum denied this request, which was accompanied by a request to exhibit them in the Royal College of Surgeons in London.

The denial was based on a report by Tomás Villanova, then professor of zoology at the Museum. He cited the extreme fragility of the specimen, "all [of whose] bones are calcined to such an extent that most of the apophyses will break at the mere touch of a hand.... I myself have noticed a number of anatomical defects in the skeleton and many of these were mentioned by Cuvier in his work on fossils, and in spite of the fact that their existence is to my discredit, I do not dare to correct them, because simply taking it apart, which would be necessary, would cause the loss of many bones and make a unique example of its own species useless."

So we are dealing with a specimen so delicate and fragile, that it may dissolve 'at the mere touch of a hand' (the fossils from the Pampas Formation of Argentina are very often described similarly in the older literature, namely as being extremely fragile and often disintegrating almost spontaneously when only exposed to the open air, see Podgorny 2012). And the Spaniards managed not only to excavate that giant but extremely delicate skeleton, without any notable major damage, then transport it all the way to Buenos Aires, mount it there, dismantle it again, ship it across the entire Atlantic Ocean, transport it to Madrid, and mount it there a second time? How on earth did they do it? At a time when modern excavation methods were allegedly non-existent? These are nowadays ubiquitous in palaeontological fieldwork, and the most important part is hardening and stabilizing the more often than not, usually extremely brittle fossil bones, and treating them with the greatest care before being taken out of the bedrock and transported to a scientific institution where they can be carefully prepared and reassembled.

To me, the story of the excavation and transport of the Madrid *Megatherium* skeleton, with the crude and unsophisticated methods of palaeontological fieldwork allegedly available at the time, borders on the brink of complete and utter disbelief. For something like that to be achieved, with such a stunning result as the spectacular mounted skeleton still on display in Madrid, you need many years, rather decades, of field experience in the excavation and conservation of large fossil vertebrates. It is completely unthinkable that a specimen like that could just be ripped out of the ground and jumbled into some crates, then go all the way to Madrid without completely disintegrating, unless the Spaniards employed methods comparable to those we see nowadays. This, however, necessarily implicates that they knew exactly what they were doing, and this, again, points to a long tradition of doing exactly that. A tradition about which we know, officially, nothing except the random account of the 1766 discovery.

Spain, in 1787, was still technically a superpower the like of which the world had probably not seen since the heydays of Genghis Khan. Although the vast Empire had for quite a while been cracking at the seams and was in constant financial trouble, it was still one of the few in the world's history above which the sun literally never set. Spain controlled the largest part of the Americas, including Central America and much of the Caribbean, as well as vast stretches of land in Africa and Asia.

No other power in the world at the time could even remotely compete with Spain, certainly not France, who had yet to gain most of its later vast colonial holdings, nor England, which had just lost many of its important North American colonies and was struggling to establish a new overseas Empire at the time. Portugal and the Netherlands probably were the only European countries which at least remotely approached the status of Spain in the late eighteenth century, but even their vast territories dwindled in comparison to the gargantuan masses of land and the uncounted millions of people under Spanish rule.

With the French Revolution and its aftermath, all that changed dramatically. Spain painfully and slowly lost all of its American colonies, one after the other, while the country itself was largely incapable of doing anything, being occupied by the French from 1808 to 1814, and, after that, being caught in constant internal struggles and economic problems. The last major overseas holdings of Spain were lost in 1898 in the American-Spanish war, and after that, the country deteriorated even more, up to the point of the devastating Spanish Civil War in the

1930s, which resulted in the decade-long fascist dictatorship of Generalissimo Franco, and after that a slow and painful adaptation to the standards of 'liberal Western democracy'.

Up to the present day, the once glorious Kingdom of Spain is a struggling country compared to other major European powers like Germany, France or the United Kingdom, with rampant poverty and unemployment and even a very unstable national identity, as the continuous struggles of the Basques and Catalonians for independence have shown in recent decades. It is very much a mere shadow of its former self, and the final nail in the coffin of the once glorious Spanish Empire was certainly the civilization reset that occurred in the early nineteenth century. Spain never recovered from its utter humiliation during that time period, and it certainly ceased to be a major player in international politics. Only one of many fundamental changes that took place during that particular time period.

But in the late eighteenth century, things were still different, and the giant *Megatherium* skeleton was finally assembled and put on display in Madrid, where it arrived in 1789, incidentally the year in which the French Revolution took place. But interestingly, not much credit went to the Spaniards for their accomplishment. Instead, Georges Cuvier published a paper on the specimen – one of his first on vertebrate palaeontology – in 1796, accompanied by two plates. His study was titled '*Notice sur le squelette d'une très–grande espèce de quadrupède inconnue jusqu'à present, trouvé au Paraguay, et déposé au cabinet d'histoire naturelle de Madrid.*' – 'Note on the skeleton of a very big species of unknown quadruped recently found at Paraguay and deposited at the Natural History cabinet at Madrid.'

As noted before, Cuvier did not even have any chance to study the original skeleton at that time (nor did he ever bother to do so ever after in his long career, a recurring theme, as we will see), but his publication of the specimen, in which he aptly named the 'unknown quadruped' as *Megatherium americanum*, thus securing the priority on the genus and species name from any potential Spanish student of the fossil, was one of the foundations of his long and outstanding career.

Just in what way did Cuvier get hold of the figures and necessary data to write this paper at the time? The Kingdom of Spain, still under rule by the House of Bourbon, which just had been not only dethroned but literally decapitated in France by the revolutionaries, was certainly not on the best possible terms with the French Republic, although it remained technically neutral. So what the fuck happened?

The *Megatherium* skeleton arrived in Madrid, neatly packed in seven boxes, on the 29th of September 1789. It was indicated that it had been found 'on the banks of the Luxan river, thirteen leagues WSW of Buenos Aires.' Work on the material was taken over by Juan Batista Bru (1740–1799) who at the time had been a long–standing employee of the Madrid museum and had already produced some outstanding work, both concerning the preparation of vertebrate skeletons and artistic illustrations.

By 1793 Bru had not only finished mounting the skeleton, but he had also produced an extremely detailed anatomical description and a number of illustrations of exquisite quality of the unique and important specimen, which were engraved as plates by Manuel Navarro. But, alas, his finished monograph was not published for reasons which to the present day are not completely clear.

A Frenchman, Monsieur Roume (Phillipe–Rose Roume de Saint–Laurent, 1743–1805), who was a representative of France at Santo Domingo at the time and passed through Madrid in 1793, got hold of the proofs of Bru's unpublished monograph. As Cuvier (1796a) indicates, another Frenchman, 'citoyen Grégoire', was the one who provided him with the plates. He forwarded them to the Institut de France, and the Insitut immediately commissioned a certain Georges Cuvier to report on the new, spectacular find.

And thus, Cuvier's account appered in 1796 in the '*Magasin Encyclopédique*' (and, in the same year, as an English translation, in the '*Monthly Magazine, and British register*'), basically ripping off Bru, who had been working for many years on the unique specimen and produced all the anatomical study and illustrations. All this, as said before, without Cuvier ever seeing or studying the original specimen. All Cuvier had were the proofs of Bru's work, which had been acquired from the Spaniards by the shady 'citoyens' Roume and Grégoire.

It is no wonder that Cuvier's original account is full of errors, among which the most prominent one is that he accidentally transferred the locality of the discovery to Paraguay. In that way, for a long time to come, *Megatherium* entered the collective consciousness of European scientists, artists and scholars as 'The great animal from Paraguay'. Although remains of *Megatherium* have since then been identified from Quaternary deposits of Paraguay, the original specimen came clearly from very close to Buenos Aires in Argentina, but even more than 30 years later it was still discussed as the '*Megatherium* of Paraguay' (e.g. Cooper 1828). So great was the influence and authority of Cuvier at the time, that contemporary scholars didn't even question an obvious blunder made by him in one of his first ever scientific publications.

But Cuvier was not the only one to publish on *Megatherium* in that year. The Danish naturalist Peter Christian Abilgaard, who, unlike Cuvier, had visited Madrid and seen the specimen first hand, also published a short note in 1796, accompanied by some rather inferior illustrations which were righteously criticized by Cuvier in 1804. And Monsieur Roume, that shady man from Santo Domingo (an interesting figure who later played a major part in the former French colony of Haiti's struggle for independence) who was so vital in bringing the proofs of Bru's paper to Cuvier's attention, published a short note on the matter himself in 1796 as well. So technically, Bru wasn't ripped off once, but three times simultaneously.

But this was not the end for unfortunate Bru. Unable to get his work published, he eventually sold his illustrations and text to a fellow Spaniard, José Garriga, who combined Bru's work with a Spanish translation of Cuvier's article, and issued it under his own name, also in 1796. Garriga – according to Cuvier (1804) a 'Captain of the Royal Spanish cosmographic engineers', not much else is known about his biography – quite heavily attacks Cuvier for his rip–off and defends Bru. Most interestingly, Garriga indicates that the mounted skeleton in Madrid was not the only specimen in the possession of the Museum at the time, but that there were at least two additional, less complete skeletons. One of them was apparently found in Peru and arrived in Madrid in 1795. The most mysterious one is the third, which, according to Cuvier (1804, p. 377) was at the time in the possession of 'le père Fernando–Scio' and apparently indeed found in Paraguay and donated by an unknown 'lady' (statement repeated by Cuvier 1804).

This is of greatest interest, as it shows that palaeontological activities by the Spaniards were not restricted to Argentina, but were conducted probably everywhere in their American colonial empire at the time. Just what became of these two additional skeletons remains a mystery. The German palaeontologists Pander and d'Alton did not find any trace of the two additional skeletons in Madrid during their visit of 1818 (Pander & d'Alton 1821) when they finally re–investigated the *Megatherium* in detail, so it can be quite safely assumed that just like so many other important fossils of the pre–Cuvier period they disappeared, likely during or shortly after the French occupation.

In 1804, Cuvier just continued to rip off poor Bru, who, at that time, was deceased without ever seeing himself credited for the outstanding work he had done. Cuvier's article "Sur le *Megatherium*", which appeared in the *Annales du Muséum National d'Histoire Naturelle*, was nothing but a blatant copy of Bru's study. As Cuvier was incapable of understanding Spanish, the French translation had to be done by Alexander von Humboldt's best buddy Aimé Bonpland, and Cuvier only added some discussion of previous works and observations of his own on the possible relationships of the animal, a subject on which Bru did not hold a strong opinion.

To Cuvier's credit it has to be said that, unlike many of his contemporaries, he correctly identified the animal not only as an edentate (the order of mammals to which, among others, armadillos and sloths belong), but also as a close relative of the extant South American sloths, whereas many others wanted to make *Megatherium* into some kind of giant armadillo, mixing its remains up with those of the true giant armadillo, *Glyptodon*, for decades. He also corrects his initial mistake and indicates that the mounted skeleton was found close to Buenos Aires (not that anyone cared).

Most importantly, Cuvier here gives at least some credit to the Spaniards Garriga and Bru, and indicates that it was Bru who mounted the skeleton, composed the plates that he reproduces, and made a detailed description, of which he presents a 'traduction abregée', an abbreviated translation. This may have been become necessary after Garriga had attacked him so fervently in his publication, and although Garriga's paper was not widely circulated and today is an absolute rarity (only two copies ever reached the USA, for example, see Boyd 1958) it painted Cuvier, deservedly so, in a very bad light. It may also have become politically opportune to give some credit to the Spaniards, as Napoleonic France and the Kingdom of Spain had entered into a short phase of improved diplomatic relationships at the time.

Cuvier also relates the story of how Roume got into the possession of the proofs of Bru's paper and how he was 'ordered' to do a preliminary description by the 'Institut de France'. Citoyen Grégoire – a certain Henri Baptiste Grégoire (1750–1831), a Catholic Priest who became a fanatic of the French Revolution – who, according to the article of 1796 got the plates to Cuvier, is, interestingly, not mentioned any more. He may have, to some degree, become an Orwellian 'unperson' in the meantime, as he was strongly opposed to Napoleon becoming French Emperor and voted against it in the French Senate in 1801.

Exactly this 1804 article – which was a 90% rip-off of the work of Bru –was later republished by Cuvier in 1812 in the first edition of his famous "Recherches sur les ossements fossiles". Not that anyone in Spain, completely occupied by Napoleon's relentless military at the time, had any reason or even possibility to complain about that.

Thus, Georges Cuvier became, in the eye of the public and the scientific community, the 'discoverer', 'describer' and name–giver of *Megatherium*, one of the most iconic fossil finds in the early days of 'modern palaeontology' while doing exactly nothing, except completely copying the work of another man and adding some 'observations' without ever studying the specimen.

So is Georges Cuvier a plagiarist? In the case of *Megatherium*, there is no way of denying it. Most of his credits are completely undeserved. The real work was done by the now almost forgotten Spaniard Juan Bautista Bru, he mounted the skeleton, he described it in loving detail, and he did all the great illustrations which were just – inferiorly – reproduced by Cuvier (and many others after him, without ever giving due credit to their creator) in 1796, 1804 and 1812.

As the study of Piñero (1988), who was the first to bring the long–forgotten achievements of Bru back to light, concludes: "A third of a century after his death, Bru had already been forgotten by the very institution where he had worked all his life. Both the skeleton and the model are preserved to this day in the Madrid Museum of Natural History, mute witnesses to Bru's otherwise forgotten labours."

Does this matter throw a particularly good light on Georges Cuvier? I think absolutely not. With his first major 'palaeontological discovery' we see him as simply a plagiarist, who reaps the fruits of other scientist's labour without a second thought and, at least initially (1796a) without giving them any of the highly deserved credit. While the Spaniard Bru is up to the present day completely forgotten by the scientific community, Georges Cuvier is, more than two centuries later, still hailed as some kind of superhuman genius who single–handedly invented the disciplines of comparative anatomy and vertebrate palaeontology.

As the case of the 'great animal from Paraguay' should have sufficiently shown, nothing could be further from the truth. All the work that Cuvier took credit for was already done before. He just had to claim it as his own, and all the world believed it and carried on, and continues to do so until the present day.

How many professional palaeontologists, I wonder, are aware of the true story of the Madrid *Megatherium*? How many still believe that it was Georges Cuvier's work that they admire when they marvel at the beautiful illustrations and detailed descriptions of the skeleton in Cuvier's publications (that is, if they even ever bother to look at them)? How many outside of palaeontological circles have ever even heard about that story?

This is one of the most blatant examples of truth stolen in plain sight in the history of palaeontology, and I think it well deserved a chapter of its own. If it was the only case of its kind, it could be forgiven as the sin of a young Cuvier who had to do just anything to become a respected and renowned scientist. The fact that he gives some credit to Bru in his later publications (from 1804 onwards) may indicate that he had second thoughts on the matter and wasn't too proud of his initial plagiarism.

But it isn't a singular event. It is one in a long series of incidents in which Cuvier did little else but plagiarize and rip off earlier researchers. It may be the most spectacular and obvious one, but it is not unusual in any way.

CHAPTER 5

The witness of the deluge – the not so dumb palaeontologists of the Scheuchzer Age and the myth of Sherlock Cuvier

One of the most blatant and widely known blunders in the history of eighteenth century palaeontology is the description of the incomplete skeleton of a giant salamander, found at Öhningen close to Lake Constance in southern Germany, as the remains of a human, perished in the deluge, by the Swiss physician and naturalist Johann Jakob Scheuchzer (1726). It is very often – and with glee – cited as damningly representative of the quality of eighteenth century palaeontology.

We have already seen that this is far from the truth. It would be about as honest as to cite the notorious case of the 'Piltdown man' as representative of twentieth century palaeontology. Both Johann Jakob Scheuchzer, the man behind the 'Homo diluvii testis', and A. S. Woodward, the one who published '*Eoanthropus dawsoni*' (as the Piltdown Man was once scientifically named) share the undeserved fate to be mainly remembered for these serious blunders which they made in their outstanding and long–lasting careers.

It is all too soon forgotten that Scheuchzer was, in his day, a pioneer of geology and palaeontology who published many works, very often beautifully illustrated, in which he described numerous fossils and was, more often than not, quite right about them. Woodward was the leading researcher on fossil fish (palaeoichthyologist) in the world at the turn from the nineteenth to the twentieth century and published numerous lavishly illustrated and highly accurate descriptions which to this day are standard references in palaeoichthyology. But of course, this is brushed aside easily, if you make one serious and, in hindsight, hilarious mistake.

The story of Scheuchzer's supposed fossil hominid, his 'witness of the deluge', has been recounted so many times that it is superfluous to tell it again. What is much more interesting here is what happened afterwards, and when and by whom the specimen was actually correctly identified. The usual story, which has made it in numerous textbooks and popular accounts of palaeontology, is well told by famous German–American dinosaur researcher Hans–Dieter Sues as recently as 2020 on the 'Smithsonian Magazine' homepage (https://www.smithsonianmag.com/smithsonian–institution/why–eighteenth–century–naturalist–believed–hed–discovered–eyewitness–biblical–flood–180973973/):

"The fossil in question is an incompletely preserved, strange skeleton that had been discovered in a limestone quarry near the small town of Öhningen in southern Germany. Scheuchzer identified his prize fossil as *Homo diluvii testis,* meaning "man, witness of the Flood." In 1726, he published a broadside to announce his discovery. In his great work on the natural history of the Bible, *Physica sacra* of 1731, Scheuchzer cited the Reverend Johann Martin Miller as expressing the hope that the "sad bony frame of an old sinner" would soften the "heart of new children of evil!"

Sues goes on to state: "Scheuchzer's interpretation of the Öhningen skeleton soon came under scrutiny. Other scholars considered the fossil the remains of a fish or a lizard. But Scheuchzer would go to his grave in 1733 convinced of his finding." But of course, our superhero, Georges Cuvier, finally comes to the rescue: "It was the great French zoologist and palaeontologist Georges Cuvier, who finally demonstrated the true identity of the "witness of the flood."

After Scheuchzer's death, the Öhningen fossil (in fact one of two salamander specimens he had in his possession) had been acquired for the collections of Teylers Museum in Haarlem in the Netherlands, where it is still exhibited today. When that country was conquered by Napoleon's army, Cuvier, then the emperor's inspector of institutions for higher education, visited Haarlem in 1811. With permission of the museum director, Cuvier examined Scheuchzer's fossil and set out to expose the bones more clearly from the soft surrounding rock. Using a sharp needle, he uncovered the shoulder girdle and both arms of the animal. With that, Cuvier established the decidedly non–human skeleton was, in fact, that of an extinct giant salamander."

So once again we have a typical story, a pre–1800 palaeontologist blunders horribly, only to be corrected by Georges Cuvier, who steps in like the older brother of Sherlock Holmes, re–examines the fossil and, voilà, we have the correct solution. The most interesting part here is what happened in-between Scheuchzer's serious mishap in 1726 and Cuvier's first investigation of the matter. This did not happen, as Sues states, in 1811 (which was the first time Cuvier saw the actual fossil, which had been acquired by the Teylers Museum in Haarlem in 1802) but already some years earlier, when Cuvier (1809a) published an article titles "*Sur quelques quadrupèdes ovipares fossiles conservés dans les schistes calcaires*" ("On some oviparous quadrupeds preserved in calcareous shales").

The contents of this article were later incorporated in the 4th volume of the first edition of his *"Recherches sur les ossements fossiles"* in 1812.

The article is very revealing. It was not long after Scheuchzer's demise, that doubts about his identification were raised. In 1758 his Swiss compatriot, Johann Gessner, declared the fossil to be that of a fish, more precisely that of a catfish (*Silurus glanis*) as it is still common in the Bodensee (Lake Constance) close to which the 'Homo diluvii testis' was found. This opinion was accepted by various subsequent authors, including the German mineralogist Rudolf Augustin Vogel (1776), the Ukrainian 'Count in exile' Grégoire de Razoumowsky (1790), and the famous German naturalist Johann Friedrich Blumenbach (as late as 1807).

Gessner's misinterpretation is much more understandable than that of Scheuchzer. As the shoulder girdle and forelimbs of the animal were still largely hidden under the sediment before Cuvier's preparation in 1811, and the pelvic girdle, hindlimbs and tail were completely missing anyway, it was quite easy to conceive the idea that one was dealing with a bony fish. And the skull of a catfish has at least a superficial resemblance to that of the giant salamander from Öhningen, certainly a much closer one than to a human skull.

Cuvier (1809) proceeds to dissect Gessner's interpretation in great detail, and compares the Öhningen skeleton to that of the catfish as well as that of an extant salamander. The verdict is completely justified: the specimen does not represent a fish, let alone a 'human skeleton' (something that no one in his right mind had believed for half a century at this point anyway), but instead an unknown form of giant salamander. The extant relatives of *Andrias scheuchzeri*, which are found in China and Japan, had not been officially known to Western scientists at that time.

Cuvier does not, however, refer to the work of another scientist, very famous in his day, namely the Dutch physician and naturalist Petrus Camper (1722–1789).

And this is where the story gets interesting. Apparently Camper already interpreted the Öhningen skeleton as that of a 'lizard' in either 1787 or 1790 (which must not mean much, as the distinction between amphibians and reptiles was not well-founded in biology at the time, and salamanders and the like were quite often simply classified as lizards; Cuvier, in 1809, classifies them as reptiles as well). I have found it extremely difficult, though, to obtain the exact reference of Camper's work where he states something about the Öhningen skeleton. Although his interpretation is mentioned even on Wikipedia, it is not referenced there, nor anywhere else, just as if it had been deliberately hidden.

I finally found a clue in the great German palaeontologist Hermann von Meyers volume *"Palaeologica"* from 1832, where a butchered reference is provided. A look through the collected works of Camper did not reveal any direct clue to this ominous paper, so it took a little more real detective's work to complete the story. In fact, we are dealing with a letter by Petrus Camper to Belgian researcher François-Xavier Burtin, which was published in 1790 in the eighth volume of the "Verhandelingen uitgegeeven door Teyler's Tweede Genootschap".

It was never published under Camper's name, though, but in fact hidden in Burtin's memoir on the 'revolutions of the Earth' on p. 35–36. Camper mainly discusses the 'fossil bird from Montmartre' described by Robert de Paul de Lamanon (1752–1787) in 1783. He comes to the conclusion that the specimen is so badly compressed that it is hard to determine, and that interpreting it as a bird would require a 'fertile imagination'.

He then goes on to cite several examples of just such fertile imagination producing strange interpretations of fossils in the past, and there he says that he considers the case analogous to "comme un lézard pétrifié a pu poser pour un anthropolithe" – "how a petrified lizard can pose as a fossil man", obviously referring to the Homo diluvii testis of Scheuchzer. So it is evident that Camper was of the opinion that Scheuchzer's specimen was neither, of course, a fossil man, nor, as many of his contemporaries thought, a fossil fish, but in fact a 'reptile'.

More important still is the contribution of Karl Friedrich Kielmeyer (1765–1844), Professor at the University of Tübingen. In an addendum to the very detailed publication of Milian Karg (1805), a physician from Constance, on Öhningen and its fossils. Karg still considers the 'Homo diluvii testis' as a catfish, including not only Scheuchzer's original specimens, but also another one that had since come to light and was at the time in the collection of the Swiss physician and naturalist Johann Conrad Ammann (1724–1812). It is now part of the collection of the London Natural History Museum. The Ammann specimen showed the extremities of the animal clearly.

It was therefore not even necessary for Cuvier to 'prepare the Scheuchzer specimen to demonstrate that Homo diluvii testis was a slamander, exposing its limbs', as conventional story-telling has it (in fact part of the limbs of the Scheuchzer specimen were already seen before, as a close investigation of the original illustrations reveals). In fact, there was already an almost complete specimen available, which showed the limbs clearly.

In the afterthoughts to Karg's (1805, p. 69–70) paper, which were added by the editor of the Journal (which by the way only saw this single volume ever published), Mr. Jaeger, we therefore read: "Mr. Professor Kielmaier in Tübingen was the first to notice that the shape and position of the extremities of this skeleton [the Ammann specimen] do not harmonize with the conditions seen in a fish skeleton, and he uttered the thought, that with respect to this, as well as the shape of the head, it has much more similarity with a crawling amphibian, and this similarity is indeed obvious to the eye, particularly when observing the scaled down figure provided here; it would be wrong to consider these extremities as bones of another animal, imprinted by chance on the same slab of shale, and not belonging to the rest of the skeleton, as they are also seen, although less clearly, in the figures of another specimen, the original of Scheuchzer (...)".

The paper also includes a very detailed and excellent figure of the Ammann skeleton, much more complete than Scheuchzer's original, and a direct comparison to the skeleton of a catfish (*Silurus glanis*). Pretty much exactly what Cuvier did in 1809 and 1812, only adding the skeleton of an extant salamander.

Cuvier credits Jaeger and Kielmeyer in both 1809 and 1812, and expresses his content that an authority like Kielmeyer 'shares his conclusions'". Well, it would probably have been much more appropriate to say that Kielmeyer came to the same conclusion long before Cuvier, and that it was published four years prior to Cuvier's first study of the Öhningen material.

Cuvier does not make any mention of Camper's interpretation of the animal as a 'lizard', which goes as far back as 1787. I would give Cuvier the benefit of the doubt here, it is probable that he just did not know (or had forgotten about) Camper's well hidden notice, although he certainly knew Burtin's monograph, which discusses a lot of vertebrate fossils known at the time in great detail. In summary, it can be said that, again, the conventional story with which we are presented is, to put it mildly, very distorted.

Yes, Scheuchzer misinterpreted the original specimen of *Andrias scheuchzeri*, and badly so. No, people did not believe that it was a 'fossil man' until Cuvier came to the rescue, no one believed it in the second half of the eighteenth century. Instead, the specimen was interpreted variously, mostly as a fossil fish of the catfish family, but already as early as 1787 as a 'reptile' (which could be either a reptile or amphibian in today's terminology) by Petrus Camper. In 1805 Jaeger and Kielmeyer already discussed the status of the animal as a giant 'crawling amphibian' (a salamander) in great detail, also providing arguments why it, in no way, could be a catfish, and provided an excellent figure of the almost complete skeleton in Ammann's collection.

Everything that was left for Cuvier to do in 1809 was to add some more data and provide a somewhat more detailed comparison with the catfish and extant salamanders, but none of it was new at the time. But, as our quote from the recent (2020) article by an outstanding authority in vertebrate palaeontology, Hans–Dieter Sues, from the Smithsonian homepage shows, everybody is still happy to perpetuate the conventional story. Sherlock Cuvier single–handedly solving the mystery of a fossil his retarded predecessors completely misinterpreted in the eighteenth century, preparation needle in hand, just like old Sherlock carries his Bruyere pipe and magnifying glass around. None of that is true, everything was known and published before Cuvier. He simply collected the data, put some of his own thoughts into the stew, and that was it.

CHAPTER 6

*Le grand animal de Maestricht
– dissecting one of Cuvier's
masterpieces*

One of the most interesting stories in all eighteenth century palaeontology revolves around the type skull of the latest Cretaceous giant marine lizard Mosasaurus hoffmanni, at the time known only as 'le grand animal de Maestricht', the great animal of Maastricht, as it did not receive its official Linnean binominal until 1829, when Gideon Algernon Mantell finally named it.

The partial skull of this Mesozoic monster is today still on display in the Muséum National d'Histoire Naturelle in Paris, although one could argue that it should not be there, as it was literally a spoil of war stolen by French soldiers when they conquered the Netherlands more than two centuries ago. The French, however, don't seem too picky when it comes to their 'national treasures' and where they originated from, be it the Obelisk that Napoleon brought back from Egypt or the giant skull of an extinct reptile that they captured from the Dutch.

A look at the first edition of Cuvier's *"Recherches sur les ossements fossiles"* from 1812, where he devotes an entire chapter to the unique find, really makes one admire the man. He provides an extremely detailed and still largely accurate description of the mosasaur, and then goes on to compare it to other vertebrates, both fossil and extant, and comes to the inevitable conclusion that it was a gigantic lizard, closely related to the monitors of today.

The study is accompanied by two beautifully executed plates, the first showing the skull, and details of it, and a comparison to the skulls and lower jaws (mandibles) of several extant lizards, the second one shows skeletal remains (some of which in fact belong to another mosasaur genus, *Plioplatecarpus*), all in lavish detail. So do we have, in this case at least, a glimpse of the 'legendary Cuvier' who single–handedly solved a palaeontological mystery that, up to that time, had puzzled palaeontologists for decades?

The type skull of *Mosasaurus hoffmanni* was found by miners 90 feet below the surface in 1780, in the marine Cretaceous sediments of the St. Pietersberg of Maastricht in the Netherlands, the type locality for the latest stage of the Cretaceous, the Maastrichtian, which makes *Mosasaurus* one of the last giant reptiles of the Mesozoic to roam the Earth – or, in that case the seas.

A widely ignored fact is, that the Paris specimen of *Mosasaurus hoffmanni*, discovered in 1780, was not the first mosasaur skull specimen to be found. Instead, the first major discovery in Maastricht was made already in 1764, but it never became really widely known, and it took apparently until 1790 before it was described in detail by the Dutch naturalist Martinus van Marum (1750–1837), interestingly in the same volume in which Petrus Camper's original identification of Scheuchzer's 'Homo diluvii testis' as a 'lizard' was published.

Cuvier tells the reader that the Paris skull later came into the possession of a certain Monsieur Goddin, who, after the city had been taken by the French army in 1794, decided to provide the specimen to the Paris Museum. This version of the story is one ad usum delphini. Another, much more colorful version, which has been widely distributed in the popular and scientific literature, goes back to Faujas de Saint–Fond (1799).

Monsieur Th. J. Goddin (or Godding) (1722–1797) was apparently far from happy to part with his prize specimen, which he himself had acquired in a quite nasty way. The specimen, so we are told, was originally examined, and the excavations directed by Johann Leonard Hoffmann (1710–1782), a local medical doctor after whom the giant saurian was later named. However, the ground below which the specimen was found was in Goddin's possession, and he started a lawsuit against Hoffmann, which he won, taking both the specimen and letting his adversary pay for the costs of the trial.

In 1794 the French army laid siege to the city of Maastricht, defended by troops of the Holy Roman Empire, but the soldiers were ordered to spare Goddin's estate at any cost, because of the already world–famous fossil. Goddin suspected that this wasn't done out of sheer benevolence, and proceeded to hide the specimen away as best as he could. The French found it a day later, the searching process was probably sped up considerably because a very 'French' reward of 600 bottles of the finest wine was promised to the lucky finders, who turned out to be a dozen French grenadiers (Faujas de Saint–Fonds 1799, Russell 1967). Poor Goddin was comforted by a sum of money and an exemption from war taxes for his loss.

The specimen and many others were brought to Paris at the command of Barthélemy Faujas de Saint–Fond, who had been commissioned by the Revolutionary government as a 'Commissary of Sciences for Belgium and the Northern Army'. Faujas produced a voluminous and lavishly illustrated monograph on the spoils of war from Goddin's collection, as well as some others (including the entire famous Stadholder's collection of William V. of Orange–Nassau), that the French 'collected' in the Netherlands. Concerning the *Mosasaurus* skull, he was of the opinion that the critter was most probably a crocodile, and he is also the one who tells the half–romantic, half–comedic tale about how the French acquired the skull.

Only that it is probably not very true. As a recent study by several French researchers (Pieters et al. 2012) found out, most of Faujas' account is a falsification, or at least a grave distortion of what actually happened, and he wasn't even present when the events occurred, but arrived two months later, so he could not witness them himself. What seems to have occurred was much less adventurous and romantic, but instead an act of outright brutality and war–crime.

According to a letter of Goddin's only heiress, what happened was that several armed French soldiers turned up at the estate of Goddin "with a carriage to collect the 'petrified crocodile' by force of arms at Godding's country house, acting under orders of Frécine" [A. L. Frécine, 1751–1804, appointed 'representative of the people' who carried out the decree of the Convention Nationale to confiscate objects of scientific and artistic interest in the occupied territories and bring them to the Paris Museum]. Pieters et al. (2012) conclude that it was "patriotism as well as his great fancy for story telling that induced Faujas to falsify the facts." He just did not want to outright explain that the fossil was brutally confiscated and brought to Paris as a war booty.

Of course, it was Georges Cuvier, the celebrated master palaeontologist of revolutionary France, who got to study the prize specimen in much more detail, and he produced his description in 1808 (again, he incorporated his former study in his *"Recherches sur les Ossements fossiles"* four years later). And more often than not, of course, Cuvier is credited with the correct identification of the remains. Whereas previous authors made all kinds of suggestions – the animal being a whale, a crocodile or even a giant fish – he came up with the correct solution. Only that, again, this is absolutely not the case, and any palaeontologist familiar with early mosasaur literature (so maybe, apart from myself, three dozen people in the world) is fully aware of that.

It wasn't Cuvier, but Adrien Camper, the son of Petrus Camper whom we just met in the previous chapter, who provided the correct identification. Adrièn Camper and Georges Cuvier had already corresponded on mosasaurs long before the French army took over Maastricht (A. Camper 1790, 1791). Camper continued to stay in contact with Cuvier, despite the 'hostile takeover', and in 1800 sent him an extensive letter, which was published in the 51st volume of the *"Journal de Physique, de Chimie, d'Histoire Naturelle et des Arts"*.

There Camper describes and discusses in detail not so much Hoffmann's skull, but a number of specimens, mostly vertebrae, which he kept in his private collection (and partly inherited from his father, who had already studied them but failed to provide a correct identification) as well as additional cranial material, including the nice skull specimen in the Teyler Museum in Haarlem, the real 'first mosasaur skull', that van Marum (1790) had described.

Camper comes to the inevitable conclusion that the fossils are "du squelette d'une espèce inconnue de reptile saurien" (A. Camper 1800, p. 279), "from the skeleton of an unknown species of saurian reptile", and he makes it very clear in the following text that with 'saurian' he specifically means some type of lizard, and not a salamander or crocodile or any other kind of critter.

He also provides two beautifully executed plates, in which he compares the lower jaw of the Haarlem mosasaur in extensive detail to the lower jaws of numerous extant reptiles, including crocodiles, turtles and lizards, and in that way elegantly and excellently demonstrates that the animal is definitely most closely related to living lizards and very different from other living reptiles.

So, once again, we see Cuvier just adding to the work of earlier authors. Adrien Camper basically already did all the work for him, and even kindly wrote him the 'letter' which became a veritable little monograph on the anatomy of mosasaurs, so that Cuvier basically had all the data at hand. As Lingham–Soliar (1995) pointed out, the main addition of Cuvier to Camper's work was that he misinterpreted the vertebral column of *Mosasaurus* and gave it a much too long neck, whereas Camper was completely right.

So does Cuvier at least give due credit to Camper in his study on *Mosasaurus*? Yes, he does indeed. He puts great praise on Camper for being the only one of the previous authors to correctly recognize the systematic affinities of the animal, and he finds himself in perfect agreement with him. Despite this, with Cuvier admitting himself in a lengthy paragraph that Camper already did the major work and reached the correct conclusions, Cuvier was the one credited with the correct identification of the *Mosasaurus* in many popular and not so popular accounts.

This is just another case of Cuvier getting all the praise for something that he did not do. All he did, again, was to compile the data of previous researchers, add some observations and illustrations of his own, and basically declare that he 'agreed' to what his predecessors had to say.

CHAPTER 7

The rock finger of Mannheim – Cuvier at his best

One of the most spectacular fossil finds of the eighteenth century was the first complete and well–preserved skeleton of a Mesozoic flying reptile (pterosaur), and it figures prominently in the early career of Georges Cuvier. The specimen, found in the late Jurassic lithographic limestones of Bavaria probably around 1780 (Tischlinger 2020), found its way to the Natural Cabinet of Mannheim, then under Bavarian rule, where it was kept until 1802 when it was removed and brought to Munich, the capital of Bavaria.

The Bavarians obviously were not interested in losing so spectacular a fossil, and they were well aware of what had just happened a few years earlier to the 'grand animal de Maestricht'. But only due to the intervention of the German naturalist Friedrich von Moll and his French friend, the famous mineralogist René–Just Haüy, was the Munich collection spared to be taken to Paris by the 'Commissaires du gouvernement Français en Allemagne', who were tasked with 'selecting' (i.e. stealing) rare books, works of art and scientific objects.

Cuvier already had an article in which he mentioned the specimen, published in 1801. As with the *Megatherium*, this was, once again, an exercise in armchair palaeontology. He had not seen the original specimen, nor did he ever bother to look at it ever even later in his career, when Munich, just like Madrid, where the *Megatherium* skeleton was kept, was under French occupation, and it would have been no problem for him to access the fossil and study it first hand. Both the detailed re–examinations of the Madrid *Megatherium* and the Collini *Pterodactylus* were exclusively and excellently done by German scientists (von Soemmering 1812, Pander & d'Alton 1821).

The story of the skeleton, later to become the holotype of the first ever named pterosaur, *Pterodactylus antiquus*, does not start with Cuvier, however, and this, at least, is widely known in palaeontological circles. The first, very extensive description, accompanied by an excellent illustration of the specimen, was published by Cosimo Alessandro Collini in 1784.

Collini, born 1727 in Florence, Italy, was a very erudite scholar. He even once was the personal secretary of the famous French philosopher Voltaire, who is widely recognized as the 'father of enlightenment' and a spiritual predecessor of the ideas of the American and French revolutions, whom he met in Berlin. This did not hamper Collini from becoming employed in 1760 as a secretary of Kurfürst Carl Theodor, then the ruler of the city of Mannheim and the adjacent countries, known as the Kurpfalz, as well as Bavaria from 1777 onwards.

Karl Theodor was a rather 'modern' monarch, though. He founded the Palatine Academy of Sciences in Mannheim in 1763, banned torture in 1776 and was highly successful in pushing the economy of his little country and attracting all kinds of scholars, artists and entrepreneurs to it. Obviously, there was also an interest in fossils at the time, and Collini was made the director of the newly founded Natural History Cabinet in 1764.

With the revolutionary and Napoleonic wars, where a large part of the former territories of Karl Theodor were incorporated directly into France, and the rest occupied by French troops, it became impossible to uphold the natural cabinet. Karl Theodor died in 1799, and most of the specimens, including the pterosaur, were removed to Munich over time. Some of them also ended up in Karlsruhe, where they are still preserved in the Badenian State Museum of Natural History. By 1803, the Cabinet was gone. Collini, heartbroken to see his life's work destroyed, died in Mannheim three years later.

In 1784, five years prior to the French Revolution and its devastating aftermath, everything was still swell in Mannheim, though, and Collini published his famous monograph "*Sur quelques zoolithes du Cabinet d'histoire naturelle de S. A. S. E. Palatine et de Bavière, à Mannheim.*" Collini's paper is often referenced but, I assume, rarely read. In fact, Collini actually does not devote a lot of space to the *Pterodactylus* fossil, but he provides a neat description, which Cuvier (1809b) quotes in full, while also reproducing Collini's illustration.

This beautiful illustration of the skeleton, which was done by the famous Mannheim engraver Egid Verhelst (the younger) (1733–1804), is probably one of the most widely reproduced palaeontological illustrations of all time. It depicts the exquisitely beautiful skeleton almost completely correctly and with all necessary anatomical detail. It seems that Collini and Verhelst very much understood what they saw.

The largest part of Collini's work is devoted to the lengthy discussion of various Quaternary mammal finds (mainly rhinoceros and hyaena remains, including complete skulls) of the kind often found and illustrated in eighteenth century works, simply because these things are just all over the place and not really rare at all.

Quaternary and Tertiary sediments are the youngest, and therefore cover the largest part of the globe. Big mammal bones are common in many of these sediments, and particularly in Europe, where remains of the ice–age fauna are found basically in every cave and gravel–pit. It is thus no wonder that these fossils, also usually enclosed in very soft sediments and not requiring too much sophisticated preparation, play such a major role in early vertebrate palaeontology.

Mesozoic, let alone Palaeozoic vertebrate sites are much rarer, the rocks usually much harder and the animals much stranger than what is found in Cenozoic rocks, and therefore, with few exceptions, these fossils were, at least by the official record, severely understudied even at the time of Cuvier. Collini, however, also discusses an ill–preserved Lower Jurassic crocodile skull from the famous locality of Altdorf in Franconia (nowadays Bavaria), so his work is in fact outstanding for depicting and describing two very early finds of Jurassic reptiles.

So, as said above, Cuvier (1809b) devoted an extended paper on the subject, and after first telling his readers that Collini was certainly well–meaning, but had only 'limited knowledge of natural history and comparative anatomy' (Cuvier 1809b, p. 424), goes on to discuss the lithographic limestones of Bavaria in some detail, before verbatim repeating the entire description of Collini. This, together with the illustration he copied from Collini's work, is almost half of the entire article.

The German pterosaur expert Helmut Tischlinger (2020) considers Collini's description as "very detailed and exact", and I can only concur to his view. Collini certainly was a capable anatomist, and he also was perfectly right when he could not place *Pterodactylus* among any existing group of animals, because it obviously represented one not found any more on Earth, a pterosaur, a group of archosaurs so highly derived and so different from, for example, an extant crocodile that, if alive today, they would certainly have been put in a class of vertebrates of their own, just like another group of flying archosaurs, the birds.

Collini also clearly recognizes that the animal had a foldable wing, and he speculates that it supported a membrane. That he regards the animal as a marine creature nonetheless is a bit disappointing, as he clearly notices its many similarities to both birds and bats. But he does not see his conclusion as definite, and basically uses the cop–out of any palaeontologist of today: 'we need more data before we can make definite statements'.

Cuvier's (1809b) article continues to correct some inaccuracies in Collini's description (mostly justified, but he also adds some errors of his own) and then enters into a lengthy discussion of the affinities of the animal. He cites Prof. Jean Hermann, of Strasbourg University, who had long wanted to 'write a memoir on the animal', and who was the one who initially pointed out Collini's work to Cuvier ('he aroused my attention in the animal', Cuvier 1809b, p. 434).

Hermann reached the opinion that the animal was 'intermediate between mammals and birds'. More importantly, Hermann was also the first one to attempt a restoration of *Pterodactylus*, which he sent to Cuvier. This restoration, long thought to be lost, has recently resurfaced (Taquet & Padian 2004), showing a flying, bat–like, furry animal with an extensive wing–membrane. Hermann sent the restoration to Cuvier already in 1800, one year before Cuvier first commented on the fossil (Cuvier 1801). Interestingly, Cuvier never published it.

Taquet and Padian (2004) also elegantly demonstrate, that Collini, who Cuvier characterized as not the sharpest knife in the drawer when it came to natural history in his 1809 article, had in fact a deep influence on Cuvier's general thoughts: "But Collini's paper also immediately provided a strong theoretical underpinning for Cuvier's prospectus for his great work, Ossemens Fossiles des Quadrupèdes. Shortly after reading

Collini's paper, Cuvier developed the introduction to his 1800 prospectus, incorporating many of Collini's rhetorical strategies with his own views. These same ideas also became important in Cuvier's Discours préliminaire to the Ossemens Fossiles, his most renowned philosophical work."

What remains Cuvier's intellectual property here is, that he correctly identifies *Pterodactylus* as neither a bird nor mammal or something in-between, but as a true 'reptile'. He provides a lot of anatomical arguments, from all parts of the skeleton, including the jaw articulation, the morphology of the hand and foot skeleton and particularly that of the pelvic girdle (which he didn't really get right himself), which he recognises as most similar to that of extant crocodiles (which it in fact is).

That the elongated fourth finger of the hand supported a membrane was, however, already speculated upon by Collini (1784) and it is explicitly drawn as a bat–like wing by Hermann in 1800, so Cuvier was certainly not the first one to recognise this correctly. Most importantly, maybe, Cuvier dares to provide the animal with a name, but it is not, as commonly believed, Ptero–dactyle (the wing finger) but actually Petro-dactyle (meaning rock finger) in the title of the 1809 article. Cuvier corrected this three years later in the first edition of the *Ossements fossiles*.

Although it is therefore very arguable whether this non–latinized and rather colloquial name, that originally had a completely different meaning, and which was finally latinized into *Pterodactylus* by Lorenz Oken in 1818, should be technically regarded as a valid generic designation, it has become common practice to ascribe the name *Pterodactylus* to Cuvier, 1809. It thus became the first genus of extinct reptile from the age of dinosaurs ever to get a generic designation.

In 1819, Cuvier at last bothered to give the animal a proper Linnean binominal, calling it *Pterodactylus longirostris*. But a specific name had already been given prior to that, by the German Researcher Samuel Thomas Ritter von Soemmering (1755–1830), a pupil of Petrus Camper. Soemmering named the specimen *Ornithocephalus antiquus* in 1812 when he published a spectacularly detailed and accurate re–description of the skeleton, so the official name today is *Pterodactylus antiquus* (von Soemmering, 1812), the specific name of von Soemmering and the generic name of Cuvier being considered the valid ones.

What is the most striking about the 'rock–finger from Mannheim' is the fact, that such an excellent skeleton was present there at the time. It is still preserved in good condition in the Munich collection today, and the recent investigation by Tischlinger (2020, p. 29) has demonstrated that it is indeed an outstandingly well–preserved specimen, he concluded "the Collini–*Pterodactylus* without doubt still counts among the best known pterosaur finds from the Solnhofen limestones which, together with its historical relevance, makes it an icon of pterosaurology".

So once again, as with the Madrid *Megatherium*, we have a completely isolated find (well, not that isolated, as we have learned already in the case of *Megatherium*) of absolutely outstanding quality, and nothing else until much later in the nineteenth century, when slowly additional pterosaur specimens came to light, first also from the Solnhofen area in Bavaria, later from lower Jurassic deposits in England and Southern Germany. The Solnhofen limestones, however, have been quarried since ancient times (allegedly the Romans already knew about them), and fossils from these localities were known and collected for centuries (Barthel 1978).

Pterosaurs are the most common fossil reptiles in these deposits, and since Collini's specimen, several hundred more or less complete skeletons and numerous isolated bones have been found and are now stored in museums all over the world, representing apart from *Pterodactylus*, at least 12 additional genera, making it the most diverse fauna of late Jurassic pterosaurs in the world. Nonetheless, not a single additional specimen was announced until 1817 by von Soemmering. His '*Ornithocephalus brevirostris*' later turned out to be a baby *Pterodactylus* (Wellnhofer 2008).

Were there really no additional discoveries ever made, or is it that only the Collini specimen is remembered today? It is a well–known fact (that no one talks about) that many early fossil collections were never properly catalogued or described (or, if such catalogues and descriptions existed, we do not have them any more), that they were sold or re–distributed after their owners died, and that very probably a large portion of early palaeontological collection and research activity is now lost completely.

There are also no reliable records of any other pterosaur finds from the eighteenth century, although an anonymous article – most probably written by D. S. Erskine, see Delair & Sarjeant 2002 – in the 27th volume of the 'Gentleman's Magazine' refers to the discovery of 'bird bones'in the Stonesfield Slate (Martill 2010), a Middle Jurassic Formation in England famous for having produced the first ever named dinosaur, the large predator *Megalosaurus*, as well as some of the first Mesozoic mammals ever discovered.

There are also some rare finds of pterosaurs from the Stonesfield Slate, but, as far as our current palaeontological knowledge goes, there were no birds around in the Middle Jurassic. The fragmentary and delicate bones of pterosaurs could easily have been mistaken for those of birds. This still happened many times in the twentieth century, and it is therefore likely that this account, which dates from 1757, indeed refers to a pterosaur discovery that predates Collini's publication of the Mannheim *Pterodactylus* by more than a quarter of a century.

Another, even more convincing indication of previous pterosaur finds was most recently published by German pterosaur and Solnhofen expert Helmut Tischlinger (2023), who unearthed a German illustration from 1759 which clearly depicts a variety of common fossils from the Solnhofen limestones, including clearly identifiable, common forms like ammonites, bony fish and fossil crabs.

In the lower left corner of the beautifully executed 'Eichstätter fürstbischöflicher Hofkalender'– the Eichstätt prince–bishops court–calendar – a huge copper engraving, 1.6 meters in height, there is a depiction of a bird–like animal in a slab of lithographic limestone. With its very long and slender beak and a short tail, it is most probably a representative of *Pterodactylus*, the most common short–tailed pterosaur in the Solnhofen Formation.

The interesting thing is, that this specimen must have been a complete skeleton, and that from its posture it clearly is not identical to the Collini specimen, and it predates Collini's work by 25 years (Tischlinger 2023)! How many other discoveries may have been there, which were never documented?

CHAPTER 8

The godfather of extinction

A very central contribution by Georges Cuvier to the emergence of the 'new science' of vertebrate palaeontology in the nineteenth century that is almost exclusively attributed to him is the concept of extinction. The conventional story is, that early researchers on fossils just did not recognize the fact that they were dealing with were the remains of animals that just did not exist on this planet any more, that had gone completely extinct a long time before the first man was there to see them. Cuvier himself claims in his autobiography that the central paper for the development of this concept was his 1796 paper 'Mémoir sur les espèces d'éléphans vivants et fossiles.' (Cuvier 1796b).

As so many others of his works, this one saw several editions and publications in various forms (many also saw numerous translations into other languages), so that even with the aid of Smith's (1993) painstakingly detailed bibliography of Cuvier, it is nigh to impossible to find a completely satisfactory way through the jungle. The first extract of the paper was published in 1796 in the *'Magasin encyclopédique'*. The complete version appeared (probably) in 1798 as a separate publication published by Baudoin in Paris, and also in the second volume of the *'Mémoires de l'Institut national de sciences et arts. Sciences mathématiques et physiques'* (given incorrectly, as '*Académie des sciences, Mémoires*' by Smith 1993) in 1799.

The final version of the article is quite detailed, and accompanied by five excellent plates of extant and fossil elephant skulls and teeth, including the Asian (*Elephas maximus*) and African (*Loxodonta africana*) elephant, as well as the Woolly mammoth (*Mammuthus primigenius*). Cuvier (1799, p. 4) states that basically all previous authors "from Tenzelius up to Pallas" had considered "the fossil bones of large quadrupeds which have been found as fossils in almost all countries of the globe [a very interesting statement, considering how little was allegedly known at the time]" as just representing the remains of the same elephants as those found today, which just happened to be more widely distributed in former times.

This is a complete straw man argument. That the American Mastodon was a species different from those living today was commonly known at the time. It had been formally named by Robert Kerr as *Elephas americanus* as early as 1792 (today it is known as *Mammut americanum*, as it represents a genus clearly distinct from the living Indian elephant).

Cuvier goes on to discuss the species of living elephants. To this end, he particularly uses skulls from the famous Stadhouder collection of Wilhelm V., Prince of Orange–Nassau, which had been stolen, together with the type skull of *Mosasaurus*, from the Dutch in 1795. He comes to the conclusion that the African and Asian elephants are clearly distinct, and that no additional species can be proven to exist at the present state of knowledge. *Elephas maximus*, the Asiatic elephant, had received its name already by Carl von Linné in 1758. That the African elephant was a different species was formalized as late as 1797 by German naturalist Johann Friedrich Blumenbach (1752–1840). A third species, *Loxodonta cyclotis*, the African forest elephant, was discovered as one of the last large living land–mammals on

Earth as late as 1900.

In the second part of his treatise, Cuvier deals with extinct elephants. His main focus is the Siberian mammoth. Known for a long time at that point and described and figured in detail by many authors, particularly the Prussian naturalist Peter Simon Pallas (1741–1811), who had at that time worked in Russia for many decades. Pallas, so Cuvier assures us, is of the opinion that the Siberian mammoth is completely identical to living elephants. Cuvier, with only a few fragmentary specimens at his disposal and otherwise relying on the literature, demonstrates a considerable number of anatomical differences, particularly in the dentition, and also shows that a number of finds from Western Europe are specifically identical with those described by Pallas and others from Siberia.

Concerning the findings from America, the Mastodons, he comes to the conclusion that they are elephants, but generically different from both the mammoth and the living forms, particularly due to the very distinctive structure of their teeth. He cites the name *Elephas americanus*, which he attributes to Pennant (probably unaware of the work of Kerr).

He finally concludes that the Siberian mammoth and the fossil elephants from America, as well as many other fossils (not to the least the 'grand animal de Maestricht', which here he still cites as a crocodile) represent extinct species, which are no longer present on this Earth, and proposes to call them *Elephas americanus* and *Elephas mammonteus*. The name *Elephas* (now *Mammuthus*) *primigenius*, had been coined, again by Blumenbach, some months earlier in 1799 in the 6th edition of his celebrated '*Handbuch der Naturgeschichte*' ('Handbook of Natural History') and is now generally accepted as the species name for the Siberian mammoth, taking priority over Cuvier's name (Cuvier also had to add a note, referring to Blumenbach's work, in the 1799 edition of his paper).

So, if Cuvier should be credited at all with being 'the scientist who demonstrated that extinct species exist', then certainly Blumenbach, who pointed out many of the anatomical differences in the dentition of mammoths and elephants just as Cuvier did, should be credited as well. But, of course, the story does not end here, because fossil vertebrates had been identified as distinct species and received Linnean binominals prior to 1796.

One of the most important examples here is the recognition of the European Cave Bear (*Ursus spelaeus*) as a distinctive and extinct species by the German palaeontologist Johann Christian Rosenmüller (1771–1820), which, as Kempe et al. (2005) state, was a "milestone for palaeontology" but is "not well known in the mainstream palaeontological literature". Cave bear bones had been known for ages in Europe, and had been recognizably figured as early as the seventeenth century.

The German naturalist Johann Friedrich Esper (1732–1781) had already published an extensive and well–illustrated monograph on Cave Bear bones found in present–day Bavaria in 1774. Esper clearly recognized that the bones could not be from the common European Brown bear, *Ursus arctos* (at that time still present in the German fauna), and suggested that instead they may be those of polar bears (*Ursus maritimus*).

Twenty years later, Rosenmüller (1794) re–investigated Esper's collection for his doctoral Thesis at the University of Leipzig, particularly focussing on a single, very well–preserved skull from Burggailenreuth. Rosenmüller compares the specimen in detail to the skulls of both the Brown Bear (*Ursus arctos*) and the Polar Bear (*Ursus maritimus*) and comes to the conclusion, that it does not represent either of them, but instead belongs to another, now extinct species of the bear genus, which he names *Ursus spelaeus*, the Cave Bear.

Rosenmüller's dissertation precedes the first ever publication of Cuvier's elephant paper by two years. Rosenmüller even proposes sexual dimorphism among the cave bear material, identifying the larger and more robust skulls as males, the smaller and more gracile ones as females, an outstanding achievement for the time, whereas later (and much more renowned) authors (including, for some time, Cuvier himself) tried to separate the two morphs into two different species.

Another important case is the 1796 monograph of the spectacularly preserved Eocene fishes of Monte Bolca in the Veronese Mountains by Giovanni Serafino Volta (1862–1842), an Italian priest and naturalist, born in Pavia (at that time still part of the Holy Roman Empire) who described more than 100 species in his contribution, of which he identified 15 (including two new genera, *Blochius* and *Monopterus*, both erected by Volta in 1796) as forms which could not be identified with any extant taxa.

Although he struggled to ally his fossil fishes with those still present in the Earth's oceans, and known at the time, the conclusion was inevitable for him that at least some of them were completely new to science (and possibly extinct, although he preferred to regard them as exotic forms the extant relatives of which had just not yet been discovered).

Like so many other important European fossil collections, the major part of the collection of Monte Bolca fossils of the Count Giovanni Batista Gazola, studied by Volta in amazing detail, was also seized by the French under Napoleon in 1797, after the conquest of Verona, and 'repatriated' to Paris, where they were later restudied by the French palaeontologist de Blainville (1818) and the Swiss pioneer of palaeoichthyology, Agassiz (1833–1844), who tried to rob Volta of the priority of all his new taxa (except one) and gave all of them new names, something fortunately rectified since then (Gaudant 2011).

The Gazola collection consisted of more than 1000 specimens – about 600 of the best ones stolen by the French – and certainly was one of the largest and most valuable collections of fossil vertebrates and other fossils from a single locality in existence in the late eighteenth century, at least among those of which we still know.

We have already commented on Kerr's (1792) introduction of *Elephas americanus* for the American Mastodon, which only was a consequence of the views held by several (but certainly not all) scientists at the time that the animal was of a kind not to be found on present day Earth.

As von Murr (1774) already writes in his report on Hunter's (1769) account on the Ohio mastodons, the 'American incognitum': "It has been a mistake of most naturalists to believe that all species of animals still exist as they did at the beginning of the world." Hunter (1769) himself already declared: "In the last place, it may be observed, that as the incognitum of America has been proved to have been an animal different from the Elephant, and probably the same as the Mammouth of Siberia; and as grinder teeth like those of America have been dug up in various other parts of the world; it should seem to follow, that the incognitum in former times has been a very general inhabitant of the globe. And if this animal indeed was carnivorous, which I believe can not be doubted, though we as philosophers may regret it, as men we cannot but thank Heaven that its whole generation is probably extinct." You cannot get more explicit than that.

Many other scientists, including, most famously, American president Thomas Jefferson (himself a very capable palaeontologist, see Osborn 1935) indeed assumed that species like the American Mastodon or the 'giant lion of the West' which Jefferson had himself discovered (and which was, in fact, the first specimen of the giant ground sloth *Megalonyx*, a North

American cousin of *Megatherium*) may still exist somewhere out in the wilds of the as yet still largely unexplored continent. But that was not the only opinion.

The rabbit hole goes even deeper than that. In his 1774 paper, von Murr refers to an even earlier detailed treatment of the question: "The great Bilfinger dealt in detail with the mammoth bones in oratione de Anatomia Elephanti, et Ossibus Mamontaeis, which is found in the 2nd fasc. of his Varior (Stuttgard 1743.8) p. 191:208."

Georg Bernhard Bilfinger (1693-1750) was, at his time, a renowned theologist, philosopher, mathematician and architect. In 1725, Peter the Great invited him to the Russian Academy of Science at St. Petersburg, as one of many German researchers. He stayed there until 1731, becoming a professor of theology at the University of Tübingen afterwards. Bilfinger was born in Cannstadt, where in 1700 numerous teeth and bones of elephants – and the skull cap of an ice-age man – had been discovered, an excavation that he witnessed as a young boy (Bilfinger 1743). Some of the specimens are still preserved at the Staatliches Museum für Naturkunde, Stuttgart.

Georges Cuvier saw the collection as a teenage boy, when he was studying in Stuttgart, and it may have been a major influence on his decision to devote a large part of his adult life to palaeontology. When Bilfinger was in Russia, the bones of Siberian mammoths started to become all the rage. And he vividly remembered the findings from Cannstadt as well.

So what exactly does Bilfinger say? His talk, "De anatomia Elephantis et ossibus Mammontaeis", was given in front of the Russian Academy of Science in 1727, but it only appeared in print, as far as I am aware, much later in 1743. It is almost never referred to in the literature. Although Bernhard Ziegler, former professor of palaeontology at Stuttgart University and head of the Stuttgart State Museum of Natural History, refers to him in a semi-popular book (Ziegler 1986), you won't find anything if you search "Bilfinger mammoth" on Google, and no relevant results on Google Scholar. There is only the Latin original text, no translation of any kind is available. Even Claudine Cohen's (2002) in-depth study on the history of mammoth research does not mention him with a single word!

Bilfinger, after discussing aspects of modern elephant anatomy, goes on to discuss the bones of the "Mamont", as he calls it, recently discovered in Siberia. The German scientist Daniel Gottlieb Messerschmidt had just returned from an expedition to Siberia, where he was sent by Peter the Great, and returned with, amongst many other discoveries, impressive mammoth fossils, including an almost complete skull, which were eventually sent to Danzig, where the German anatomist Johan Philip Breyne studied them and published on them in 1741.

Cohen (2002) considers Breyne's publication as "the "official" eighteenth-century document identifying the mammoth as an elephant." Bilfinger, in 1727, does with no word doubt that these are true fossils of elephants. It should be remembered that only a few decades earlier, when the fossils of Cannstadt were excavated, many scientists were still of the conviction that they were formed spontaneously by some 'vis plastica' in the earth, or at least that is the official story.

Bilfinger's study precedes Breyne's publication by 14 years! It is noteworthy that Breyne, who deems all the mammoth bones as the remains of animals which were killed by the deluge, quite in contrast to Bilfinger's much more thoughtful interpretation, does not make any mention of Bilfinger in his publication.

Bilfinger does not waste any space on such ridiculous ideas. He compares the Cannstadt finds and those from Siberia, notes similarities and differences, and deems them all to be genuine elephant fossils. Then he goes on and dissects various other, less retarded hypotheses that were widely discussed by his predecessors and contemporaries.

One, namely that the elephant bones were remains of animals that the Romans brought with them, he dismisses on the valid ground that there is no historical record of the Romans ever bringing African elephants to Germany. "Desunt autem Historici, qui deductos in Germaniam a Romanis elephantem memorent", "There are no historians who mention the elephant being brought to Germany by the Romans." Nor, he states, is it likely that the 'Emperors of Tartaria' ever brought Asiatic elephants to Siberia.

He also completely dismisses the idea, entertained at the time by Johann Jakob Scheuchzer and many others (the 'diluvianists') that these fossils were the results of a single, world-wide flooding event, the grand deluge of Noah that the Old Testament tells us about. He argues against this from various lines of reasoning. His statement "Cautius agi velim et castius, cum sacra ad explicationes physicas documenta transferuntur" "I would like the transfer of sacred rites to evidence of physical developments to be treated more cautiously and more chastely", is remarkable.

He concludes that under the assumption that the animals lived in the regions where they are found today, there is no reason to assume that they were the victims of a world-wide flood. As reasons for their demise and burials, he assumes earthquakes and regional floods, something much more in line with modern concepts of taphonomy, and also miles ahead of Cuvier's cataclysm theory, developed more than three quarters of a century later! He also explicitly mentions, although in passing, that the various stratigraphic layers which yield fossils can hardly be explained by a single world-wide flood.

These are already astounding insights for an early eighteenth century scholar which, according to the official 'mythology of science', should be nigh impossible. But it gets even better than that. After dismissing the prevalent ideas about the origin of the elephant fossils, Bilfinger presents his own conclusions on the matter. In his opinion, the elephants of today, used to lavish food sources and warm climate, are not representative of the entire elephant genus. These animals once also lived under harsh conditions in Germany and Siberia They had to adapt to these harsher conditions, tougher food and cold, and so only the most resilient survived, but they did not propagate as much as those in the warmer regions. The animals perished in various ways, were buried when they were covered by landslides during search for food, when they retreated into caves or when they drowned in rivers.

We have here indications, and more than subtle ones, that Bilfinger in 1727 entertained ideas that were more than a century ahead of his time, including adaptation, survival of the fittest and speciation. His text should be considered a major milestone in the history of palaeontology and found in all textbooks on the subject. Instead, it is completely memory-holed. It is self-evident that he considered these 'cold-weather elephants' extinct, as they simply did not exist any more in either Germany or Siberia.

It is certainly noteworthy that Bilfinger was an admirer of the great German philosopher and mathematician Gottfried Wilhelm Leibnitz, who in his 'Protogaea' already noted: "Others wonder at the species one sees everywhere in stones, for which you would seek vainly in the known world, or at least in our local places. Thus, they say that Ammon's horns, which many consider a kind of nautilus, sometimes differ in form and size (for some have been found that are a foot in diameter) from all other creatures found in the sea. But who has thoroughly explored the ocean's secret recesses and subterranean abysses? How many previously unknown animals did the New World give us? It is also conceivable that many animal species were transformed by these great upheavals." (cited after Cohen

& Wakefield 2008). Here we also find, although cautiously expressed as possibilities, the concepts of both extinction and the transformation of species. Leibnitz wrote his Protogaea between 1691 and 1693!

In this context, it is noteworthy that the first definite identification of the mammoth as a fossil elephant still can't be credited to Bilfinger. Already in 1696 the German scholar Wilhelm Ernst Tentzel published a study on a fossil elephant skeleton, excavated at Burgtonna, Thuringia in 1695 (it was not exactly a mammoth but a specimen of the straight-tusked elephant *Palaeoloxodon antiquus*). With the help of the study of the Irish anatomist Mullen (1692) on the anatomy of extant elephants, he clearly demonstrates, much to the distress of his peers, who insist on the thing being a fossil unicorn, an 'unicornu fossile', that it is a fossil elephant (Gaudant 2010).

Tentzel (1698) again published his results in the Philosophical Transactions of the Royal Society two years later. That the famous Italian scholar Scarammucci (1697) had come to his aid and supported his identification sure strengthened his confidence to do so. Tentzel sent some samples of the elephant skeleton to the Royal Society, which was not only pleased to receive them, but also supported his identification.

Scaramucci's (1697) assessment was certainly influenced by the fact, that even previously the Italian scholar Ciampini (1688) already had identified fossil bones found near Vitorchiano in Lombardy as those of an elephant. This research became known to the German scientific community as it was published in a German Journal (edited in Nuremberg) in the form of a favourable review article by Hieronymus Ambrosius Langenmantel of the University of Tübingen (Langenmantel 1689). None of these authors entertained the idea that the elephants in question were an extinct species, though.

So, in summa Cuvier was again certainly not the first here,

and he was definitely preceded by Rosenmüller in 1794, but Rosenmüller's studies never gained the recognition that they deserved and everybody and his dog has forever been citing Cuvier's elephant study as the holy grail of the concept of extinction – or rather the first real elegant proof of it, based on large extinct vertebrates. This is, like so many other myths surrounding Cuvier, simply not true, and it is about time that scientists who preceded him and did excellent work in the eighteenth century, such as Rosenmüller, Volta, Bilfinger and Hunter, finally get the recognition they deserve.

Once again here, we also see a marked bias of official palaeontological historiography in favour of certain personalities, most prominently Cuvier, but also in favour of and against certain countries. As Cohen (2002) puts it: "The history of palaeontology has also been told as the story of the heroes of science. Since Cuvier, an enduring myth exists of the palaeontologist as a new hero, "un antiquaire d'une nouvelle espèce," who is able to decipher the vanished monuments of the past, to dig them out of the layers of the earth and spectacularly reconstruct monstrous animals from small pieces of bones, thus giving new life to entire extinct faunas. Making the history of science a gallery of great names—in this case the names of Cuvier, d'Orbigny, Gaudry, Boucher de Perthes, and Boule in France; Darwin, Falconer, Owen, and Huxley in England; Leidy, Marsh, Cope, and Agassiz in the United States—would follow this heroic trend."

Revolutionary and post-revolutionary France, England, the United States. The main power-engines behind the 'Great Reset' of the late eighteenth and early nineteenth century, the beacons of 'enlightenment'. We see this over and over. No mention of Spain. No mention of the Holy Roman Empire, or Germany in general. No mention of Italy or Russia. These were all, at the time, still 'backward' nations which had yet to be fully baptised through bloody wars and revolutions and made part of the new emerging world order. They are not allowed by official history to have played a major role in the chain of events which led to the establishment of the 'new science' for a 'new world. And we have seen, time and time again, that this is not true, that scientists from Germany and Spain in particular made major contributions that are almost completely memory-holed. The cases of Bilfinger and Rosenmüller are just two in this extensive list.

CHAPTER 9

The little girl and the Great
Sea Dragons – palaeo-
feminism gone wrong

E very good story needs a heroine as much as a hero, and while France received its poster–boy in the shape of Georges Cuvier, England had its poster–girl instead, little Mary Anning (21 May 1799 – 9 March 1847), popularly credited to this day as the discoverer of Ichthyosaurus, Plesiosaurus and the first British pterosaur, later to be named Dimorphodon macronyx.

Mary Anning is a figure right out of a Charles Dickens novel, and she was the only woman who made any kind of noteworthy contribution to the field of vertebrate palaeontology in the early nineteenth century, albeit only as a collector and outright fossil dealer. In consequence, as the science they want to do has been exclusively created by old grumpy white men otherwise, Mary Anning has become an icon for palaeontologists with a feminist inclination worldwide.

Certainly, more books have been written on her than about any of the men who did the real scientific work on the fossils she discovered, and she even has become the protagonist of fictional novels. She certainly is not "The unsung hero of palaeontology", as the homepage of the Natural History Museum in London (https://www.nhm.ac.uk/discover/mary-canning–unsung–hero.html) just called her in an article from 2022, of course written by a lady by the name of Marie–Claire Eylott.

Miss Eylott seems to busy herself as a 'Digital Content editor', whatever that may be, I never heard that she had made any achievements in palaeontology. Nonetheless, the once mighty and unfuckwithable British Museum features her article prominently. Of course, "The first ichthyosaur" is attributed to Mary Anning in that article, as well as the first complete *Plesiosaurus* and the first pterosaur found outside Germany.

All of this is questionable on many levels, the ichthyosaur story, of course, is complete bullshit and I consider it hilarious that this retarded myth is perpetuated by the London Natural History Museum on its official homepage in the year 2022. But it is just a very telling demonstration that in the fake world of today, truth does not matter any more, not even in so–called 'science'.

Far from being 'the unsung hero of palaeontology' (shouldn't it be 'heroine' or would that be politically incorrect? What were her 'pronouns'?), Mary Anning has been stuffed down the throat of everybody ever even remotely interested in palaeontology since kindergarten age by thousands of popular and scientific articles, TV programs, books and other media, so that quite to the contrary she is probably among all the official pioneers of vertebrate palaeontology the most widely known to the public.

As Noè et al. (2019) recently put it: "Mary Anning (...) is perhaps the single most famous historical fossil collector in the world, about whom more has been written than almost any palaeontological figure". Yes, we know she was there. Yes, we know that she did not have a dick. We don't need 1000 more books and TV shows and unfunny memes about her to tell us the fact, thank you.

Far from being 'an unsung an oh–so–suppressed and oh–so–forgotten and oh–so–female pioneer hero (or heroine)', as the always politically correct Natural History Museum laments, she has become a quasi–mythical figure, as well as an idol for wannabe palaeontologist girls worldwide. "Princess of palaeontology and geological lioness". No, she did not get bestowed with such titles in the politically correct twenty-first century, but already in the nineteenth century by her peers, including one Gideon Algernon Mantell, the totally unimportant guy who discovered *Iguanodon* (Davis 2012)! That Mantell also described her as 'a prim, pedantic vinegar looking female; shrewd, and rather satirical in her conversation' may be a bit less convenient to her fans. So, just in case that you have any doubts that she served as a poster–girl already in her lifetime, please take notes.

The contributions of her have been overblown to an extent that is unprecedented, with the possible exception of Monsieur Cuvier himself. In the place our world has become, it is pretty much unthinkable that a hero of science doesn't have a female counterpart. And as there wasn't a single female scientist in sight during the early days of vertebrate palaeontology, the fossil dealer Mary Anning had to fill that gap. Isn't it funny how the people at the Natural History Museum are always on the forefront when it comes to bashing those shady and evil fossil dealers, yet of course it is all different and oh–so cute and oh–so great when a girl does it? It is probably politically very incorrect to question the true significance of Mary Anning's contributions, but I do it anyway, because political correctness has nothing to do with the truth. Also, it is great fun.

A lot of myths surround her life, starting with the unlikely story how she was struck by lightning as a little girl, survived and became exceptionally bright afterwards, while she had been a pretty stupid brat before. I consider that story as fishy and part of the mythology that was deliberately built around her, already during her lifetime, but I am not interested here in dissecting her biography. It has been commented upon ad nauseam.

What I want to investigate is the popular myth that she 'discovered' the Great Sea Dragons of the Jurassic, Thomas Hawkins' 'Gedolim Taninim of Moses', the famed ichthyosaurs and plesiosaurs.

So let's start with the ichthyosaurs, a group I incidentally know a bit about. Mark Evans (2010) gave what he probably considered a quite comprehensive overview of early ichthyosaur (and other fossil reptile) discoveries and had to say some words on the matter. Suffice it to say at this point that his article isn't even remotely comprehensive at all. If one reads his paper, one gets the impression that all early ichthyosaur discoveries were made exclusively in England. In fact, most of the important early ones were made in Germany, but these aren't mentioned with a single word in his paper.

This leaves two possibilities, which are equally unpleasant for the author. Either he is a flaming Anglo–Saxon chauvinist, or he is completely ignorant about a large part of the field in which he allegedly is an expert and has just more or less copied everything from earlier accounts (like that of Howe et al. 1981) without doing any research whatsoever on his own part. We will come back to the German finds and their significance later. Let us stay in England for a while.

Evans (2010) notes that there are records of a '40 foot long young whale' discovered at Weston near Bath in the year 1766. This, very likely, was an ichthyosaur skeleton, as the beasts indeed resemble dolphins and whales the most among living creatures, and Liassic plesiosaurs don't get that big. If it indeed was 40 feet long, it must have been one of the giant forms, most probably a *Temnodontosaurus platyodon*. Exactly the same species which started the Ichthyosaurus craze and was 'found' by Mary Anning, then 12 years old, in 1811.

Just that it wasn't found by her, but according to all existing accounts (even the Natural History Museum hast to admit it in its article), by her older brother, Joseph. The Anning's pretty much ran a more or less lucrative family business, involving Mary's mother and various of their children. And there was obviously no concern about sending the young kids out to the dangerous coastline to do the hard work. Child labour was a thing in early industrial revolution England, why should it be any different for a fossil dealer family?

Of course, we know the drill by now, no one knows what became of the 40 - feet ichthyosaur. There is no figure, no description, no account about whether it was even collected or left on the shore to rot, nothing. The only reason why we even know about it is a private letter that Torrens (1979) luckily unearthed. It would have been the most spectacular find of a fossil reptile in the entire eighteenth century, and even the biggest specimens nowadays in the great Natural History Museums would possibly pale in comparison to that giant.

And all we have is a private letter. I consider this a fact that is worth more than a fleeting thought. So we are made to believe that right in the middle of the eighteenth century, the heydays of the Age of Enlightenment, when fossil collecting became a big thing among the princes and gentlemen all over Europe, a 40 feet long (!!!) saurian skeleton was discovered close to one of the towns most popular with the high society of England and that it just disappeared? That nobody gave a rat's ass about it? Obviously, yes, as that is what happens to pre–Cuvierian (or in that case pre–Anningian) discoveries on a regular basis, as we have already seen. If we have lost American dinosaurs, lost German pterosaurs, lost Peruvian and Paraguayan Megatheriums, why not a lost 40-foot British ichthyosaur?

But the monster from Bath was not the only discovery. An ichthyosaur skull from the famous locality of Barrow–on–Soar apparently entered Cambridge University in 1799. The skull, again, is completely gone, only a plaster cast of it seems to remain (Evans 2010). Howe et al. (1981) note the numerous ichthyosaur specimens – most from England, some from Germany – present in the Hunterian museum, assembled by famous and ill–fated anatomist John Hunter (1728–1793) and painstakingly catalogued by Richard Owen. 29 of these were already present in Hunter's time, so right in the middle of the eighteenth century, before Mary Anning was even born. Ichthyosaur fossils apparently were neither rare – they are, in fact, the most common fossil reptiles in both England and Germany by far – nor unknown to English scientists and collectors during the eighteenth century.

It would have been interesting to know what Hunter thought about these specimens, but as most of his manuscripts remained unpublished and were later deliberately destroyed by the shady Sir Everard Home (yes, exactly the man who described the 'first Ichthyosaur' of Mary, excuse me, Joseph Anning), who very likely plagiarized a lot of Hunter's discoveries, we will never know.

Hunter's remaining notes, of course, tell us that he considered the vertebrae as fish, the teeth as those of sea lions. A pretty dumb and typically 'pre–Cuvierian' interpretation. Haha, the old dumbass, he didn't know better. Need some enlightened nineteenth century chap to come up with the real solution. And his girly sidekick, of course.

What Hunter's notes destroyed by Home said, we will never know, but it is probably not out of sheer coincidence that Home became so interested in the Anning ichthyosaur. As Mark Evans notes, citing the work of Flower (1898): "Flower believed that if Hunter's researches, based on specimens in his museum, had been published they would have elevated him to the pioneering status now held by figures such as Cuvier." Maybe, just maybe, Hunter had second thoughts on his ichthyosaurs?

But the evidence has been completely and utterly erased by a well–meaning fellow gentleman and scientist, who was 'Hunterian professor' and in charge of the 'Hunterian Museum' at the Royal College of Surgeons, the highly decorated and nobilitated Sir Everard Home. That Hunter's collection was largely destroyed by a German bombing raid in 1941 wasn't too helpful as well. Life has a sense of irony sometimes.

But let's leave the British ichthyosaurs for a while. Certainly, the most famous ichthyosaur localities in the world are not in Britain, but in the vicinity of Holzmaden and Bad Boll in Southern Germany. That numerous fossils occur there in the Lower Jurassic Posidonia Shales had been known for an eternity, Johannes Bauhin had already figured numerous fossils from there in excellent quality in 1598! So was nothing found there in all the time? And did no one ever bother to look at it?

As hard as it is to believe, but the first scientific description of finds from southern Germany is usually attributed to Georg Friedrich von Jaeger (1785–1866), who published his monograph "De Ichthyosauri sive proteosauri, fossilis, speciminibus in agro Bollensis in Wurtembergia repertis" in 1824, a long time after the ichthyosaur craze had started in Britain. He figures quite a bunch of material on several lithographic plates.

Among them is a specimen that was collected already in 1749 by the German physician Christian Albert Mohr (1709–1789). In contrast to many British finds that mysteriously disappeared, the skeleton, surviving Napoleon and two world wars, is miraculously still well–preserved and on exhibit in the State

Museum of Natural History at Stuttgart. Its skull and much of the tail are missing, but it is a perfectly articulated specimen otherwise. And, what is the most important, it shows something that the Holzmaden area has become famous for more than a century later: it is the skeleton of a pregnant female, with the skeleton of the embryo still inside!

Yes, that is right, more than half a century before Mary Anning's (excuse me, Joseph Anning's) 'first ichthyosaur' there was a much more spectacular specimen from southern Germany. Why on earth did it take until 1824, when Georg Friedrich von Jaeger came up with the first detailed description (incidentally and unfortunately one of the last in vertebrate palaeontology to be completely written in Latin, so that I assume it is rarely read, and even more rarely understood, by today's students)? Were all German researchers for almost a century too dumb or to idle to do anything with it?

Mohr left a manuscript, to which von Jaeger (1824, 1828) refers in his later treatises on fossil reptiles, mentioning in passing that he regarded the critters as 'cartilaginous fish' or 'rays'. But unfortunately, it was never published. Certainly Mohr's specimen is definitely, much more than the Anning specimen, the 'first complete ichthyosaur ever discovered'. Even though mama ichthyosaur misses the head, it is completely preserved in the embryo, so that, apart from the tip of the tail, the entire skeleton is present in this 'buy one, get one free' ichthyo.

Why isn't this spectacular early discovery not much more widely known than the pretty much run–off–the–mill *Temnodontosaurus platyodon* discovered by the Anning's? Maybe a grumpy old German physician isn't as sexy as a British teenage girl? Of course, the Natural History Museum doesn't make a single mention of the Mohr specimen. Neither does the article on Anning on the English and German Wikipedia.

But was the Mohr specimen a single find? That the Posidonia

shales were rich in fossils were known at least since the late 16th century (Bauhin's monograph). The spectacular pseudoplanktonic crinoids had been beautifully figured and described in great detail by the German theologist and hobby-scientist Eberhard Friedrich Hiemer (1682–1727) already in 1724! Fossil crocodiles in the form of Cuvier's 'Gaviale de Boll', later officially named *Crocodilus bollensis* by Georg Friedrich von Jaeger in 1828, had been known as well for a long time. The type specimen, an incomplete postcranial skeleton, was acquired by the Dresden Museum already in 1730!

Quenstedt (1856) indicates, that several complete skeletons were collected by Mohr and transferred from his collection to the Königliches Gymnasium at Stuttgart, later ending up in the Königlich Württembergisches Naturalienkabinett, which today is the Stuttgart Natural History Museum. But he also laments (Quenstedt 1856, p. 217) that all these early efforts had been practically forgotten, until Home described the Anning specimen in 1814!

Eberhard Fraas (1891, p. 3) after giving an extremely brief and incomplete review of early discoveries, mostly isolated vertebrae, which does not even mention the Mohr embryo specimen, indicates: "It is understandable that with the state of the natural sciences in Germany at that time, no further conclusions could be drawn from such isolated finds, and a correct interpretation was only arrived at after entire skeletons had been found in England and described by competent naturalists." This is bullshit.

As Quenstedt (1856) indicated decades earlier, several at least subcomplete skeletons were at the disposal of German scientists from Mohr's collection alone. And even the single embryo specimen, which showed the entire ichthyosaurian osteology, would have been more than sufficient to arrive at a sound conclusion about the animal. That for so many decades nobody ever seems to have studied these specimens in any detail is a complete and utter mystery.

And there must have been more specimens, probably plenty of them. Rebstock (1720) mentions a 'skeleton with ribs, spine and vertebrae', weighing about 50 kg, that had been taken to a collection in Tübingen. Keyßler (1740–41) mentions fossil skeletons from Bad Boll, one of which was in the collection of Dr. Mauchardt at Tübingen (possibly the Rebstock specimen?), while another one (obviously the latter type of the marine crocodile *Macrospondylus bollensis*) went to the collections at Dresden, to which it was sold by the Tübingen Professor and fossil collector Gmelin in 1730 for the hefty sum of 500 Thaler.

Apparently the fossil dealing got so out of hand in these early days, that the Count of Württemberg decided in 1749 that it was necessary to keep an eye on these things. So much for Mary Anning as the first big professional fossil dealer. To that purpose, Christian Albert Mohr, medical doctor at Göppingen, not far from the Bad Boll/Holzmaden area, was assigned (and paid 50 Gulden each year) to keep an eye on new fossil finds and, if possible, collect them and integrate them in the collection of the Stuttgart Gymnasium. The first time, to my humble knowledge, we see some kind of 'fossil protection decree' enacted anywhere in the world. Only two of Mohr's specimen, a small incomplete skeleton and the big mama with the embryo inside, are still identifiable in the Stuttgart collection. But according to Quenstedt (1856) there must have been plenty more.

So, to recapitulate: by the middle of the eighteenth century the occurrence of fossil reptile skeletons, both marine crocodiles and ichthyosaurs, at Bad Boll was widely known. And not only that, but there was even already an extensive trade in these fossils far beyond the borders of small Württemberg, so extensive, in fact, that the count himself got annoyed by the treasures of his territory disappearing. Some of these specimens, like the Mohr double–whopper, were as complete as the Anning ichthyosaur.

It remains a mystery why no scientific literature on these finds existed prior to 1824. Were the German scientists really as retarded as Fraas (1891) assumed, needing the help of brilliant and benevolent British gentlemen like Home and Conybeare to finally come up with something decent of their own? Or are we missing something here, just like Wistar's first description of a North American dinosaur, just like Hunter's destroyed manuscripts? How can we even be sure that the literature we have at our disposal is complete? What about Mohr's manuscript, mentioned by von Jaeger (1824, 1828)? I have never seen it. Has there been more like that? We will probably never know.

But enough of the ichthyosaur shenanigans, what about plesiosaurs? Particularly what later became the holotype skeleton of the arch–plesiosaur *Plesiosaurus dolichodeirus*, painstakingly excavated by young Mary Anning all on her own, with primitive tools and no scientific support, over a period of ten years, a real tear–jerker? And probably her most famous discovery. Sure, she will always be remembered as the girl who discovered *Plesiosaurus*, you cannot take that away from her.

Well, you can, if you are a nasty guy like me. Kind of. I have occasionally done some research on sauropterygians, the reptile group that includes plesiosaurs, so I feel mildly competent to comment here. Sorry to say that plesiosaurs were already know a century before Anning's discovery, and not by too shabby specimens as well. The most interesting of these early discoveries is another specimen nowadays referred to *Plesiosaurus dolichodeirus*, not from Lyme Regis, where the Anning plesiosaur was discovered, but from Fulbeck in Nottinghamshire.

It was first described by William Stukely (1687–1765) in the

Philosophical Transactions of the Royal Society in 1719. Yes, that's right, more than 100 years before the Anning specimen, which was found in December 1823 and described by Conybeare (1824). And it wasn't just a few vertebrae that Stukely was dealing with. In fact, it is almost half a skeleton, comprising much of the trunk, parts of both hind fins and part of the tail. And it belonged to the great–grandfather of one Charles Darwin originally.

That is probably the reason why it hasn't got 'lost', like so many other important pre–Cuvierian finds, but is still well–preserved and even on display in the Natural History Museum of London. The Anning narrative is sure important, but the Darwin narrative beats it by a long stretch, so everything connected to Saint Darwin is of course sacrosanct.

What did Stukely make of the find? Originally regarded as 'a human skeleton', he rather feels that the thing is a crocodile or porpoise. Well, as plesiosaurs are, in some way, looking like a cross–over between the two, he isn't that far off the mark. Stukely also determines the majority of the preserved bones correctly.

He also mentions another interesting (and apparently completely lost) early palaeontological find in comparison: "a Crocodile, tho' a small one, found after the like manner inclos'd in Stone, from a Quarry in the Mountains of Upper Germany." I wonder what that thing was and where it ended up!

He then adds a lengthy discussion on the nature of fossils and comes to the conclusion that his critter was an "amphibious or marine animal" and that it was preserved because "this Sceleton, with others like it, fell entire into the Fissures of this Bed of Clay, which has since turn'd into stone". That is a pretty decent taphonomic statement for early eighteenth century standards, so Stukely does a fine job.

But the Darwin plesiosaur wasn't the only one known in the eighteenth century. The first volume of Nichols' (1795) monograph on "The History and Antiquities of the County of Leicester", describes and figures a virtual multitude of definite plesiosaur specimens, including a partial skeleton, from the local Leicestershire private collections of Reverends Mounsey and Turner.

The pretty decent figures show plesiosaurian dorsal, cervical and sacral vertebrae, plesiosaurian limb bones and two articulated specimens, the larger one definitely a plesiosaur! A practically complete skeleton, most probably of a marine reptile and possibly a plesiosaur, in Turner's collection is mentioned (Nichols 1795, p. CCV): "In the collection of the Rev. Mr. Turner, rector of Denton, in the vicinity of Belvoir castle, are two blocks of stone, hard but coarse, and of dull–greyish colour; these, when in one mass, about four or five feet long and nearly two broad, were luckily split, and discovered the convex and concave skeleton of an animal; the bones were not much distorted from their natural position; they are completely petrified and are become of the same nature with the stone itself."

Any determination of that intriguing find except that it is some kind of 'quadruped', so certainly not a fish and most probably not an extremely fish–like ichthyosaur, is wanting. None of the material from the Mounsey and Turner collections is accounted for. Like so many other pre–Cuvierian finds of great importance, this tremendous and unique early collection is completely lost.

So there may have been a complete (or almost so) plesiosaur in Rev. Turner's collection, there definitely was a partial skeleton in that of Rev. Mounsey, and the Darwin's had the Stukely specimen at home in the eighteenth century already. I don't think I have to come up with more evidence. Yes, Mary Anning found the first complete plesiosaur officially known, and it was the first one to attract the attention of scientists competent enough to determine what it was that she found, at least as far as we know, but it was of course not the first plesiosaur specimen, and neither was it even the first articulated skeleton that was found. They had been known at that time for more than a century!

And what about pterosaurs? Yet another group of fossil reptiles I've worked on at times, so let's do the notorious 'fact–checking'. Mary Anning is variously credited in popular and not–so popular accounts, via her 1828 discovery of the first specimen of the Lower Jurassic pterosaur *Dimorphodon macronyx*, as having discovered the first pterosaur outside of Germany – and therewith the first British pterosaur, the first Lower Jurassic pterosaur, and the first long–tailed pterosaur (what we used to call rhamphorhynchoid pterosaurs). Sorry to burst the bubble of the world–wide Mary Anning fan club once again, but none of this is correct.

The first substantial find of a rhamphorhynchoid that I am aware of was again made in the Solnhofen Limestones of Germany, in 1825, so three years prior to Anning's discovery. The specimen later became the type skull of *Rhamphoryhnchus muensteri*, described by Goldfuss in 1831. This is widely known among pterosaur researchers and needs no further comment.

The first definite British pterosaur remains were mentioned by Gideon Algernon Mantell in 1824, incidentally his first ever published paper on palaeontology. The fragmentary stuff from the Wealden (Lower Cretaceous) was originally interpreted by him as a Cretaceous bird, which he later named *Palaeornis cliftii*. Most probably it is an indeterminate azhdarchoid pterosaur (Martill 2010), so this discovery predates Anning's *Dimorphodon* find by four years. We already mentioned the specimens from the Stonesfield Slate discovered even earlier, which were most probably also pterosaurian in nature. These may be doubtful, the Mantell pterosaur is not.

So did she discover the first Liassic pterosaur, at least? Officially, quite certainly. The first specimen of *Dimorphodon macronyx*, described by William Buckland in 1829, was found by her in 1828, and there is no definite surviving record of an earlier pterosaur find from Lower Jurassic strata. The important word here is: surviving.

Buckland (1829, p. 219) notes: "I had for some time past suspected the existence of the Pterodactyle in the Lias at Lyme; partly from having heard, about twenty years ago, that in the collection of Mr. Rowe, then made at Charmouth, there was the skeleton of a fossil bird, which I never saw, but imagine may have been a Pterodactyle; and partly from having found, four years ago at Lyme, in the collection of Miss Philpots, some bones of a wing and toe, which I could refer to no other animal, and of which a drawing was then made for me. More recently, I have discovered in the cabinet of Miss Philpots a thin elongated fragment of flat bone, which appears to be the jaw of a Pterodactyle; it is set with very minute, flat, lancet–shaped teeth, bearing the character of a lacertine animal— A drawing of it is annexed." Food for thought? I guess so.

And then, of course, there are the finds from the Posidonienschiefer Formation of the vicinity of Banz in Franconia, Bavaria, which Carl von Theodori first published at the end of 1830. Theodori already lists nine specimens, all of them isolated bones except for three associated wing elements, the most diagnostic of which (and later type–specimen) is an almost complete lower jaw.

Who has ever collected fossils in the Posidonia Shale Formation knows how exceedingly rare pterosaurs are, even isolated bones, so to have amassed such a substantial collection sure was an outstanding feat, and it is highly likely that at least some of these specimens were discovered prior to 1830, and that Theodori, who was slow and careful when it came to publication of his results, waited until he had enough material to make at least some reasonable statements.

Unfortunately, neither von Theodori (1830, 1852) nor more recent papers on the Banz pterosaurs of Padian & Wild (1992) and Padian (2008) make any mention of when exactly the specimens were found.

But doing your own research and actually READING the literature instead of just copying and pasting from anyone else does wonders in science. The paper of von Theodori from 1831 in Oken's 'Isis' explicitly mentions when the specimens were found, and as it is dated as being from January 1831, no doubts remain. Interestingly, one reads (von Theodori 1831, p. 277): "Three years ago, I found, on a slab from the middle layers of the Blue Lias, which was exposed at the occasion of a road construction, likely near Kleinhereth, one hour away from Banz, three complete bones and a fragment, which I mostly cleared of matrix, and which belong to the genus Pterodactylus." So the start of 1831 minus three years leaves us at the beginning of 1828! As I consider it unlikely that major road constructions were carried out or Theodori was much in the mood to collect fossils in the bitter cold Franconian winter, it is likely that the specimen was discovered in spring 1828. The remaining specimens, which von Theodori already mentions in his 1830 paper, including the nice lower jaw, were not found by him, but by his collector buddy Geyer while Theodori was busy recovering a big crinoid slab in the summer of 1830.

And when did Mary Anning recover the first *Dimorphodon* specimen? Exactly, in December 1828 (Noè et al. 2019). So, for me, the case is satisfactorily settled. The first ever specimen of a Lower Jurassic pterosaur that we still know of was not Mary Anning's *Dimorphodon*, described by Buckland in 1829, but Carl von Theodori's *Dorygnathus*, described by him in 1830. And everyone who ever really read von Theodori's original publications, instead of just citing them, would have come up with the exact same conclusion almost 200 years ago!

So what remains of the three most outstanding vertebrate palaeontological feats of Mary Anning? Did she discover THE Ichthyosaurus? No, she didn't. She discovered and excavated part of a specimen (the notorious Homeian Proteosaurus), namely the postcranial skeleton, of an ichthyosaur which became a historically highly significant exemplar, as it was vital in starting a wave of scientific research on these animals, first in England, some years later also in Germany.

Did she discover THE *Plesiosaurus*? No, but certainly more so than THE Ichthyosaurus. She discovered a very fine and complete specimen, the later holotype of *Plesiosaurus dolichodeirus*, the first ever complete skeleton (that we know of, think about the critter of Rev. Turner) that again attracted the attention of a capable scientist (William Dean Conybeare) and was the central pivot of early nineteenth century plesiosaur research.

Did she discover THE first ever long–tailed, Lower Jurassic and non–German pterosaur? No, none of that, Georg Graf zu Münster, Gideon Algernon Mantell and Carl von Theodori did, as far as it can be at present ascertained, bearing in mind how much information, how many specimens and how much of literally everything we have probably lost from the times before the early nineteenth century 'Great Reset'. What she discovered was a well–preserved articulated partial skeleton of a pterosaur that was the first generically and specifically diagnostic find from Great Britain, and which became the type of the important early pterosaur genus and species *Dimorphodon macronyx*.

All these are outstanding achievements for a lady with little school and no academic training in the difficult times of the early nineteenth century, achieved by a combination of luck as a fossil collector and untiring, hard work. But it is not necessary, and does not do any justice to her, to heap superlatives upon her, making her some kind of palaeontological 'Wonder Woman' and ascribe world records to her which are just wrong.

But as we have seen already, the early history of palaeontology is full of myths, legends, distortions and outright lies. But if such distortions, myths and lies are perpetuated for literally centuries, even by leading scientists, one is left to ponder whether this happens just out of laziness or retardation, or because of a sheer disregard for the truth in favour of some open or hidden agenda. In the case of the only noteworthy female figure in palaeontology during the time frame which interests us, the agenda is blatantly obvious, and any further explanation is superfluous.

To end this chapter, I want, with the greatest sincerity, to express my personal admiration for Mary Anning (and faithful Tray), who has made extremely important contributions to vertebrate palaeontology through her amazing discoveries at a time when any kind of science–related endeavours were still nigh to impossible for women. She was, by all means, an admirable and outright heroic lady, and it is not her fault that she was later instrumentalized and turned into a larger–than–life icon. But the truth must be told, whether we like it or not, or we will forever be caught up in the spiderwebs of a distorted and faulty historical narrative.

CHAPTER 10

The dinosaur rabbit hole

D inosaurs, as conventional palaeontological history tells us, did not play any role in the early days before the 'Great Reset'. We all know the stories of the first discoveries, and the iguanodontian adventures of Gideon Algernon Mantell and his wife. The tales about the Megalosaurus of Reverend William Buckland and the first finds in North America, including Leidy's Hadrosaurus skeleton, have been told and retold thousands of times, with only minor variations, and have become part of a collective 'mythology of science' with which generations of palaeontologist have grown up since the dinosaur books of their childhood days.

Georges Cuvier only plays a minor role in these tales, as all these early discoveries of relevance were made in England, and to England the centre of palaeontological research shifted after the defeat of Napoleon, just as it would shift to the United States in the post-Darwin years from the middle of the nineteenth century onwards. Just as it had shifted from the Holy Roman Empire to France before, not to the least because of the French conquest and occupation and the relentless stealing of the fossil treasures that had been assembled there and were relocated to Paris, where Cuvier and his colleagues were eagerly waiting to study them.

Concerning dinosaurs, usually only Cuvier's correspondence with Mantell, related to the *Iguanodon* teeth, is mentioned (and more often than not, an incorrect account is provided), but we will see that he did some more important stuff that is now almost completely forgotten.

Dinosaurs were a post–reset invention, at least in the public mind. Although very little was officially known still in the Middle of the nineteenth century, they had become extremely popular. They figured prominently in the World's Fair of London of 1851, in many ways the single most important event to start off the 'modern age of science and technology' which we have enjoyed ever since. The life–sized models by British sculptor Waterhouse Hawkins, most of them – in contrast to the spectacular Crystal Palace itself, which was destroyed by fire in 1936 – still preserved in Hyde Park, were presented to the six million visitors of the fair, at that time equalling one third of the entire population of Great Britain!

It can be said with no overexaggeration that the presentation of Hawkins' models, featuring the founding fathers of the Dinosauria, *Iguanodon*, *Megalosaurus* and *Hylaeosaurus*, as well as other Mesozoic monsters, such as *Ichthyosaurus*, *Plesiosaurus* and the giant amphibian *Labyrinthodon* (*Mastodonsaurus*), done under the guidance of Richard Owen, was the single most visited, most successful and impactful dinosaur exhibition in human history. It deeply planted the giant extinct reptiles of the Mesozoic into the collective consciousness of humanity more than anything else ever before or after, including even Spielberg's 'Jurassic Park' franchise.

A very similar exhibition was done just two years later in 1853 on the other side of the Atlantic Ocean in New York, again involving Hawkins, who produced life–sized sculptures of some of the discoveries of Cope, Leidy and their predecessors, including a *Hadrosaurus* fighting with two specimens of Cope's fierce dinosaurian predator *Laelaps* (*Dryptosaurus*). Again a Crystal Palace was involved, and again this one was destroyed by fire. As a sidenote: they all were, if they weren't deliberately demolished. And at one time, almost every big city in Europe and America had one. None of them remains. A very interesting story. Be this as it may, the American sculptures, unfortunately, were also gone after the catastrophe, and only illustrations of them survive.

The exhibition must have been just as spectacular as that in London, but it was less of a success, mainly due to the internal political problems in America in general, and corruption ridden New York, where it took place, in particular. But, at any rate, after the two great fairs in London and New York, dinosaurs were now well established as the icons of the 'new and modern' science of palaeontology on both sides of the Atlantic in both countries that mattered. Before that time, it had been the spectacular Quaternary and Tertiary mammals described by Cuvier, Blumenbach, Jefferson and others, then the sea–dragons, the ichthyosaurs and plesiosaurs described in the 1820s mainly by English scientists like de la Beche, Conybeare, Home and König.

Now it was the dinosaurs, erected as a group of their own just ten years before the London World Fair by Richard Owen, who, in many ways, had become the successor of Georges Cuvier as the main gatekeeper of palaeontological research before he was himself made largely obsolete by the advent of Darwin's theory of evolution by means of natural selection and the spectacular American discoveries of Cope and Marsh.

Owen founded his Dinosauria on very slim evidence, because not a single complete skull, let alone a skeleton was known at the time, at least officially so, and it is no wonder that the Crystal Palace models were, to our eyes, vastly inaccurate. Owen did get one thing right, however, probably one of the most important features. He presented the dinosaurs as very mammal–like in habits and posture, more like giant reptilian pachyderms than oversized lizards and crocodilians.

In this way, he was extremely modern, and his models were excellently conveying the idea that dinosaurs were physiologically and behaviourally more comparable to extant mammals and birds than to living reptiles, concepts that became mainstream only late in the twentieth century.

But there was a deep underlying reason for Owen's presentation of the dinosaurs, and why they were chosen to play such an important role in the further development of the 'new science' of vertebrate palaeontology. The dinosaurs, these reptilian behemoths of old, were animals well fitting into the new world order. They were something completely novel and unheard of in the days before the 'Great Reset'. They were huge, and strong, and powerful, just like the new steam–powered machines invented and employed everywhere during the industrial revolution, just like the new American and British Empires that emerged after the collapse of the old world from 1778 to 1820.

And Britain, as well as, after a slow start, particularly the United States of America, were evidently the leading countries in the discovery and study of these animals. New fossil giants for a new world order, and the leading powers of the new order could boast the most spectacular finds of these creatures and present them to millions of astounded plebs in giant expos and later in giant newly erected museums, like the British Museum in London, which was built with the profits from the London 1851 World's Fair and, of course, came under the directorship of none other than Richard Owen.

These giant nineteenth century museums became temples of science, and extremely important for public education (or re–education in the spirit of 'modernity' and 'enlightenment'). Particularly after Darwin, and after a big push for evolutionary theory and, by extension, social Darwinism, it became necessary to justify the extreme social injustice as well as the brutal and genocidal conquests which characterized most of the remaining nineteenth century.

It was okay that things were as shitty as they were for uncounted millions of people, because it was just the same in nature for uncounted millions of years. Survival of the fittest, and those who can't adapt are left behind and finally eradicated. The law of nature demanded it, and there was nothing wrong. This distortion of Darwin's ideas played a major role in late nineteenth century and early twentieth century politics, and it arguably does to the present day.

But back to the dinosaurs. The usual explanation – if any – for the lack of dinosaur records prior to the early 1800s is that dinosaur fossils are, as compared to those of Cenozoic mammals, more rare, usually more incomplete, more difficult to determine – as most of them are so dissimilar from anything alive today – more difficult to excavate, prepare and conserve.

Of course, there are exceptions to this, but these rules of thumb indeed hold true for the majority of discoveries and so it is expectable that researchers of the eighteenth century and earlier did not know anything about dinosaur fossils and never got to see one. Only that we already learned that they had well-preserved and spectacular fossils of other Mesozoic reptiles, like *Mosasaurus*, *Pterodactylus*, marine crocodiles, ichthyosaurs and plesiosaurs at their disposal, even several skeletons of the much older, late Permian early archosaur-relative *Protorosaurus* were known, but these are usually portrayed as 'great exceptions'.

The only pre–1800s dinosaur discovery that is sometimes mentioned in text–books and popular accounts is, again, and not by chance, a scheuchzeresque blunder. The distal (lower) end of the upper thigh-bone (femur) of a *Megalosaurus*, presumably from the Stonesfield Slate of Cornwell near Chipping Norton, in

Oxfordshire, England, was figured by Robert Plot (1640–1696), professor of chemistry and first director of the Ashmolean Museum at Oxford, as early as 1677 in his 'Natural History of Oxfordshire'. Later labelled – and allegedly interpreted – as the fossil ball sack of a giant humanoid by Robert Brookes in 1763.

As Rieppel (2021) recently made it clear – not being the first to do so – that Brookes did decidedly not consider the specimen the fossil testicles of a man, but clearly refers to them as part of a thigh-bone, following Plot's interpretation, and the labelling of the specimen was most probably done in error. Halstead (1970), with glee, pointed out that the label 'Scrotum humanum' ('human scrotum') which is found in Brookes' work, would technically be the oldest available Linnean binominal for a dinosaur.

Halstead & Sarjeant (1993), recognizing that they unopened a tiny little Pandora's box, applied to the International Commission of Zoological Nomenclature to suppress the name of Brookes as a nomen oblitum (a forgotten name), but this application failed.

Scrotum humanum is technically a valid Linnean binominal. The only reason that it is not applied instead of Megalosaurus bucklandii is the fact that it is considered to be too incomplete to be sure whether it represents this or another species of big theropod, so it is technically a nomen dubium (a dubious name), and if future research should point out characters in the end of the femur unique to Megalosaurus bucklandii among known dinosaurs, the name would take priority.

As the specimen is – as most early dinosaur finds – nowadays lost and can't be re–investigated, it is not too likely that this is ever going to happen. Alternatively – and probably best – it could be dismissed as a nomen erratum, a name applied in error (probably by Brookes' engraver) which was never meant to be a valid Linnean binominal in the first place (as advocated by Rieppel 2021, and I agree with him on that point).

But, as usual, this highly distorted story only picks out a single blunder (and a very harmless one, in that case) from the 'Scheuchzer age' the like of which are still absolutely common in the science of palaeontology and still happen today on a constant basis, even to renowned experts, when dealing with fragmentary specimens – to contrast it with the 'great advancement' made during the nineteenth century. We have seen this pattern already over and over, and we know the drill now. We also should have learned by now that it is not true at all. And the same, of course, is also the case for dinosaurs.

Dinosaur fossils were much more widely known before the nineteenth century than is usually recognized in mainstream palaeontological textbooks and popular accounts, and even their first correct interpretation may be way older than what we are made to believe.

These earliest dinosaur discoveries, as far as they are still documented and verifiable – which is often difficult, as we will see that in almost all cases the original fossils have been completely lost, just like so many other pre–1800 discoveries – are a deep rabbit hole and I will demonstrate that those few of which we are aware, mostly due to the decade–long labour of love of two British researches, Justin Delair and William Sarjeant, who spent decades on researching the topic and published numerous papers on it (the most important ones being those from 1975 and 2002), are very probably only the tip of an iceberg, and that we can quite safely assume that dinosaur fossils were widely known and scientifically discussed long before the discoveries of *Iguanodon* and *Megalosaurus*.

Delair & Sarjeant (1975, 2002) give plenty of examples of such early discoveries, which had remained unnoticed by most previous researchers. Plot's *Megalosaurus* bone fragment was not the only dinosaur specimen to be already figured in the seventeenth century. In 1699, Edward Lhuyd (1660–

1709), Plot's successor at the Ashmolean Museum, figured an unmistakable theropod tooth from the Stonesfield Slate in his *'Lythophylacii Britanii ichnographica'*, as well as what is definitely the tooth of a basal sauropod, probably some kind of *Cetiosaurus* which is common in the British Middle Jurassic. Additional fossil teeth, of sharks, bony fish and crocodiles, are also figured.

Again, all this material is lost, but it represents a nice little collection of Middle Jurassic fossil vertebrate teeth, representing numerous taxa and including at least two different genera of dinosaurs, a theropod and a sauropod. And this already in 1699! We can safely assume that the Ashmolean Museum at that time held a substantial collection of Mesozoic fossil vertebrate remains, and that maybe additional dinosaur specimens were kept there.

Again, of course, all the specimens figured by Lhuyd (as well as any others that may have existed at the time) are now completely lost, all that remains is a single account and the figures, which are executed in such a good quality that after more than 300 years it is still easily possible for a competent vertebrate palaeontologist to identify most of the figured specimens.

According to Gunther (1925), the Ashmolean Museum held a geological and palaeontological collection of more than 1700 specimens under Lhuyd. Of these, only two remained in the museum in the 1940s (Gunther 1945)!

This well–documented case should sufficiently demonstrate that my hypothesis that the largest part of early fossil collections has become lost prior to, during or after the 'Great Reset' of the late eighteenth to early nineteenth century isn't some figment of my imagination, but simply what really happened.

If even a famous fossil collection in one of the world's most renowned scientific institutions, the University of Oxford, just disappeared almost without a trace, what, do you think, has happened to innumerable collections held privately or in less well–known and well–curated places before or during these troublesome times?

A great example, also described in detail by Delair & Sarjeant (2002) is the private collection of the French Huguenot emigrant Smart Lethieullier (1701–1769). The handwritten catalogue of his collection (which also comprised some dinosaur bones from Stonesfield) is an amazing 528 pages, with numerous beautiful hand–made illustrations, many of them coloured, of some of the best specimens.

Lethieullier had among his fossils not only some dinosaur remains, but also other vertebrate specimens, including those of fish, ichthyosaurs, crocodiles and fossil mammals. All that remains of Lethieullier's collection is the singular catalogue which has survived by pure chance in manuscript form. His fossils, including the dinosaur specimens, are, like so many others, completely gone!

John Woodward (1665–1728) describes a broken thigh-bone of a large theropod dinosaur in his posthumously published catalogue of British fossils (Woodward 1728), and this specimen at least has survived and is still kept in the Woodwardian Collection of the Sedgwick Museum at the University of Cambridge.

Joshua Platt (1669–1773) also collected fossils from the Stonesfield Slate. He discovered several huge vertebrae (probably dinosaurian, never published and now lost) in 1754 and an almost complete huge well–preserved right theropod femur around 1757. The latter was published, accompanied by a decent drawing, in the *'Philosophical Transactions of the Royal Society'* (Platt 1759). Unfortunately, this excellent specimen is, again, no longer available.

Many other dinosaurian finds from the Stonesfield Slate and other localities in Britain may have already existed at the time, now lost to science. One of the few that survives is a theropod scapula, presented to the Woodwardian Museum at Cambridge by a certain Dr. Watson in 1784, which was never studied or figured up to Delair & Sarjeant's (2002) groundbreaking work.

So it is obvious that dinosaur fossils were widely known, collected, held in museums and private collections and published upon in England since the late seventeenth century. What did scientists make of them? Did anyone ever suggest reptilian affinities, or were they just generally regarded as some curiosities of nature, remains of ancient 'giants' or maybe ancient mammals?

The first author hitherto identified who refers to these bones and indicates that they are from large unknown reptiles is the famous German naturalist and mineralogist Abraham Gottlob Werner (1749–1817), the director of the Bergakademie (Mining Academy) of Freiberg in Saxony, one of the oldest institutions where geology (in a modern sense of the word) was taught to students. Werner was an immensely influential scientist in his day, and any opinion he held on geological, mineralogical or palaeontological matters was certainly recognized by his contemporaries (his book was also translated into English in 1805).

He writes: "And a petrifaction of an unknown animal, but bearing relation to the Crocodile, has been found in Aluminous–Shale among Ammonites on the sea coast near Whitby in Yorkshire, and also in Compact Limestone near Blenheim, the seat of the Duke of Marlborough." (Werner 1774).

The crocodile–like animal from Whitby is quite certainly the famous teleosaurid crocodile skeleton from the Alum Shales found there in 1758 by Captain William Chapman (and probably additional finds now lost and forgotten). As also pointed out by Delair & Sarjeant (2002), the reference to "compact limestone near Blenheim" – which is very close to Stonesfield – clearly refers to the Stonesfield Slate.

The Stonesfield Slate is indeed largely a "compact limestone" (and now officially known as the 'Taynton Limestone Formation') and not a slate. Only three thin layers within the formation were slaty limestone which was quarried in the eighteenth and nineteenth century for roof tiling, leading to the many fossil discoveries there.

So Werner in 1774 already provided a very decent interpretation of the large bones from Stonesfield, namely that they were unknown reptiles close to extant crocodiles, or, in modern classification, unknown archosaurs. Without any concept of 'Dinosauria' this was probably as close as anyone could get at the time to a perfect determination.

So already in 1774 a geologist of great importance and international renown had correctly identified the 'giant animals' from Stonesfield as some unknown, crocodile–like saurians. And this not on the basis of a single find, Brookes' 'fossil ball sack', but a considerable number of specimens which were already known at the time and some of which had been published and figured since the late seventeenth century. Please note that Werner's identification of the Stonesfield material as a giant unknown reptile predates Buckland's description of *Megalosaurus* (who did not come up with a more exact identification of his 'giant lizard', quite to the contrary), the 'first dinosaur', by half a century!

But even Buckland's (1824) famous account of *Megalosaurus* was based on much older finds. Delair & Sarjeant (2002) were still unsure when the material had been discovered and only indicated that it must have been before 1818. In fact, the iconic dentary (anterior lower jaw bone) which Buckland (1824) figured in such amazing detail and which is now considered the type–specimen of *Megalosaurus bucklandii*, was already discovered in 1797 (Gunther 1925, Howlett et al. 2017), more than a quarter of a century prior to the publication by Buckland!

Yet it was not only in England that dinosaur fossils were recovered in the eighteenth century. The best known example from France is a partial skeleton of a large theropod dinosaur from the Middle Jurassic, closely related to but distinct from the British *Megalosaurus bucklandii*. After much nomenclatural knick–knacks, it now bears the name *Streptospondylus altdorfensis*, which was first proposed by the German palaeontologist Hermann von Meyer in 1832 (Allain 2001).

The skeleton was discovered by a clergyman, Abbé Bachelet of Rouen, at the Falaises de Vaches Noires, in the vicinity of Villers–sur–Mer, on the French Atlantic coast, to the present day a famous hunting ground for fossil collectors from which a number of important fossil vertebrate specimens, as well as uncounted ammonites and other fossil invertebrates, have been recovered.

Cuvier, in a short note on the find published in 1800, indicates that the skeleton was discovered, like the rest of Abbé Bachelets collection, most probably already in the 1770s, although an exact date is, as yet, not known (Cuvier 1800b). A short note by the French naturalist Jacques François Dicquemare (1733–1789), which already appeared in 1776, may even indicate another dinosaur find at that time, an incomplete femur, but as the material is lost once again, this can't be definitely proven any more.

Unfortunately, the Abbé had not only found the partial dinosaur skeleton, but also several specimens of fossil marine crocodiles in the same deposits. All this material was mixed up in a truly horribly way by Cuvier (1812) in his chapter on fossil crocodiles in the *'Recherches sur les Ossements fossiles'* and by most later authors.

After the official discovery of the dinosaurs in Great Britain, Cuvier recognized, however, that the limb and girdle bones which were preserved were clearly not crocodilian, but much closer to the *Megalosaurus*, and figured them accordingly in the second edition of his magnum opus (Cuvier 1824). Much of the dinosaur specimen, namely all the vertebral material, continued to be considered a somewhat weird crocodile far into the nineteenth century, until the French palaeontologist Gustave Lennier finally recognized all the remains as dinosaurian in 1887.

The type material of *Streptospondylus altdorfensis*, which was probably an early relative of the Cretaceous sail–backed giant *Spinosaurus*, is particularly remarkable, because, unlike most of the finds from England, it consists of a pretty good skeleton, including numerous vertebrae of the neck and back, part of the pubic bones of the pelvis, part of the lower thigh-bones (tibia and fibula) and two ankle bones of the foot (astragalus and calcaneum), so it is by far the most complete dinosaur specimen (officially) found in the eighteenth century, certainly much more complete than the initial discoveries of *Megalosaurus* and *Iguanodon*.

That Cuvier did not recognize it for what it was, and only partially, so after the British dinosaur finds were published, once again underlines that fossil reptiles were probably not a field of research in which he really did anything amazing. Cuvier always was, first and foremost, a palaeomammalogist, as were most early researches in vertebrate palaeontology, simply because fossil reptiles formed a relatively minor part of the available material, for reasons already discussed above.

But eighteenth century dinosaurology does not end here. Apparently, there were also finds in North America that predate the famous discoveries of Leidy, Cope and Marsh by many decades. As Delair & Sarjeant (2002, p. 192) report: "The earliest North American discovery appears to have been in 1787, when a large bone, considered to be a 'thigh-bone', was discovered near Woodbury Creek, Gloucester County, New Jersey in Late Cretaceous strata. The find was reported to the American Philosophical Society in Philadelphia on October 5 of that year by Dr Caspar Wirter (1761–1818) [This should read Caspar Wistar, famous American anatomist and the man also deeply involved, among other things, with Thomas Jefferson's *Megalonyx*] and Timothy Matlack (see Hindle, 1956, p. 270). The fate of that specimen – probably a hadrosaur limb–bone – is unknown."

So again we have an important specimen discovered many years prior to the first official dinosaur discoveries in the US which, again, is mysteriously 'lost', 'misplaced' or disappears outright afterwards. It seems that Matlack and Wistar presented the specimen at a meeting of The American Philosophical Society, where both Benjamin Franklin and Georges Washington were present and had a chance to see it (Thomson 2008)! They read a paper entitled '*A large thigh bone found near Woodbury Creek in Glocester County, N.J.*' but, alas, no copy of their manuscript has ever turned up to the present day (Thomson 2008). It is as completely lost as the specimen itself!

Of course, it can not be definitely proven that this bone was that of a dinosaur, but as the Upper Cretaceous Greensands of New Jersey were basically the 'Stonesfield Slate' of early American dinosaur research and yielded the type skeletons of both Leidy's *Hadrosaurus foulkii* and Cope's *Laelaps aquilunguis* more than 70 years later, it is not too unlikely that it was indeed a dinosaur, probably a hadrosaurid, as these are the most common dinosaurs in that formation (Weishampel & Young 1996).

This story, unlike most others related to very early dinosaur finds (and most probably because it happened on American soil) is the only one possibly known to a wider public, as it figures in Bill Bryson's (2010) bestselling book "*A short story of nearly everything*". As Thomson (2008) points out, there remains uncertainty whether a large bone, equally from the Greensand of New Jersey, mentioned in the diaries of 'Mastodon–Man' Charles Wilson Peale in 1797, was the same specimen or not.

In my opinion, it is pretty clear that it was a different specimen. If Matlack and Wistar presented their report on the dinosaur bone in 1787, why would they have it lying around in some godforsaken place in rural New Jersey for another ten years? Particularly, as the American Philosophical Society urged them to immediately "search for the missing part of the skeleton" (Thomson 2008: p. 41). Was sending Peale to collect the bone for them a decade after their initial report their way of responding to the request of the American Philosophical Society? I guess that is highly unlikely.

So we can almost safely assume that the Peale specimen was a second eighteenth century dinosaur find from America. Again, nothing definite is known about its further fate. Peale, who notes that he wanted to acquire it for his private museum in Philadelphia – where a near–complete Mastodon skeleton, that after a strange odyssey finally ended up in a German museum, was the prize specimen – never mentions it again. He sold his ill–fated collection in 1850. Yet there remains a bone – a hadrosaur middle foot bone (metatarsal) with both ends missing in the collections at Philadelphia, which was noted much later by Leidy to have been from 'Peale's museum' (Thomson 2008, p. 44), so it is possible that this is the bone in question. But it can't be proven any more.

Peale's career as a collector was a long–lasting one, and it is easily possible that he acquired this specimen much later, as I consider it very unlikely that no dinosaur specimen should have been found in New Jersey for a time–span of almost 70 years after the original discovery in 1787, and several fossil bones from New Jersey found in the early nineteenth century are mentioned in the accession catalogue of the Peale Museum (Thomson 2008: p. 45).

The picture that emerges from these accounts of early discoveries clearly is, that dinosaur teeth, bones and even partial skeletons were collected, published on, widely known, and even quite properly identified already by scientists of the eighteenth century, and not only in England, but also in France and the USA (and, possibly, elsewhere, although we have no remaining knowledge about it). But most of the specimens that 'officially' existed are now lost forever. Entire collections, be they private or housed in renowned institutions, have simply disappeared without a trace.

And Cuvier? He's not allowed to play any major role in the early discovery of the dinosaurs, despite his contribution by identifying at least the limb and girdle bones of the *Streptospondylus* skeleton in 1824. He is, like the skeleton itself, usually not much commented upon, often not even mentioned in accounts of early dinosaur research. Cuvier, in contrast to many other cases we have examined here already, is treated pretty unfairly when it comes to dinosaurs and his substantial scientific contribution is not widely acknowledged.

By 1824, revolutionary France and Napoleon had served their purpose in the 'Great Reset' agenda of the nineteenth century. The Holy Roman Empire and the Spanish Empire were destroyed, the British Empire and the United States were on the rise as new global superpowers, and France had basically already handed over the baton to England at that time, which in the later half of the nineteenth century became again replaced by the USA as the major force, as it still is today. With a short intermezzo in the latest nineteenth and earliest twentieth century when, arguably a historical wild-card, the new–founded German Empire took over as the major player, as it did in record time in all other fields of science and technology for a short period, before it was defeated and destroyed by the First World War.

But the story that we are presented with, that it all started with Buckland, and Mantell and Owen, and was then continued by Leidy and Cope and Marsh and – poof – we have the dinosaurs, all in post–Napoleonic Europe, after the initial 'Great Reset' phase that lasted approximately until 1820, is obviously a vast distortion of historical truth.

It is just that we are left with very limited clues. The surviving manuscripts and specimens from the time before 1800 may only be the pitiful remains of a once much vaster body of knowledge. In the case of the collections once available to researchers, this is definitely proven now. The well–documented case of Caspar Wistar's first American dinosaur discovery (if it really was the first, which is highly doubtful) where not even the manuscript remains, indicates, that also our scientific literature from before 1800 is more incomplete than we generally assume.

Were all of these things that 'just happened', because the naturalists and museum curators and scholars of old were a bunch of forgetful retards who treated their treasures with complete neglect? Or is there something more sinister hidden here? An agenda to erase the past to such an extent as just to leave a minor part of it intact? So that a plausible timeline of events for later generations can be constructed? Letting a small portion of the previous knowledge survive to have plausible deniability and a 'foundation myth' for the explosive development of the 'new science' in the post–Napoleonic world?

To examine this a bit further, it is necessary to dive somewhat deeper into the interconnections of naturalists, researchers and scholars involved in palaeontology at the time – as well as many other sciences. It will be seen that most of the people who were active and of importance during that time period, or whose works have been promoted by later generations (most prominently those of Georges Cuvier) were deeply connected with each other on many levels, and that most of them were also strongly involved in the politics of the day.

Most of these early researchers were not living in an academic ivory tower behind the ivy walls, as the domesticated scientists of today who have been trained since early childhood not to question things, but to behave and follow the paradigm. These people were actively participating in a lot of important developments of the late eighteenth and nineteenth century, and almost all of them entertained personal friendships and intimate connections to some of the most influential figures in politics, economy and the military.

By the end of the eighteenth century, they were like a big international elite club, operating outside the standard societal institutions, and many of them were actively promoting ideas that came to fruition in the American and French revolutions. They were themselves promoting and furthering the agenda of the 'great civilization reset', and they were usually happy with the outcome. We will throw more light on these things in the next chapter.

CHAPTER 11

The Republic of Letters – it's a big club, and you're not in it!

A widely prevailing concept, not only in the history of science, but generally held by most scientists (and other intellectuals) themselves, which emerged in the days of the early Renaissance (Fumaroli 1988, van Miert 2014) is, that they formed part of a highly distinct and different class of society, practically a big elite club, the membership of which could only be attained by intellectual and scientific 'merit'.

It is usually referred to as the 'Republic of Letters', in German, the term 'Gelehrtenrepublik' ('the republic of scholars') is roughly equivalent. This 'Republic of Letters' was by the eighteenth century, the 'Age of Enlightenment' which culminated in the American and French revolutions and their aftermath, in many ways a real 'state within the state' among large parts of Europe and at least North America, particularly after the founding of the USA.

Many well–documented cases show how scholars from many countries cooperated on certain large–scale projects, like the observations of the Venus transit in 1761 and 1769. These eventually involved more than 500 scientific observers from numerous countries on a worldwide scale.

Even ongoing wars did not hamper the endeavours of the 'Republic of Letters', "Despite the war between France and England in 1761, the British Admiralty guaranteed the French astronomer Pingré safe passage to make his observations." The same in 1769: "Scientifically oriented explorers such as Captain Cook were granted immunity by all parties even in time of war; the French Ministre da la Marine went to the trouble of posting bills in all ports instructing sailors to render to Cook all necessary services" (Daston 1991, p. 378).

In the eighteenth century, the 'age of reason', scientists and intellectuals were no longer seclusive scholars, hidden behind walls of books in solitary studies. They had become a widely revered class of society, and their achievements regarded as a source of prestige comparable to winning a war. They were seen as almost on par with the traditional aristocracy (Daston 1991), and socializing with the elites of the ancien (as well as the nouvel) regime was commonplace among scholars of the time. As it was usually the kings and princes, or the presidents and dictators who paid their salaries, there was a very vital reason for doing so, but of course it also greatly increased their self-esteem.

The 'Republic of Letters' was inherently internationalist and cosmopolitan. Great academies and scientific societies were keen on getting foreign and corresponding members. The use of a common 'lingua franca', first Latin, in the eighteenth century more and more replaced by French, and by English in the later nineteenth century up to the present day, was only one of the many features by which the members of the 'Republic' set themselves apart from the common plebs.

Although several European countries, the German parts of the Holy Roman Empire (Eskildsen 2004) and Russia (Jones 2004) in particular, went their own ways to a certain extent, many in the intellectual circles of these countries – those in Russia were largely immigrants, particularly from Germany, France and the Netherlands anyway – nonetheless considered themselves as members of this cosmopolitan club.

And its members had enormous privileges that no one else enjoyed in the day. As Daston (1991, p. 375 f.) points out: "Letters flowed back and forth across the Channel between members of the Royal Society [of London] and its opposite numbers in the Académie des Sciences [in Paris] throughout the eighteenth century, despite strained diplomatic relations and outright war between the two countries; much the same can be asserted regarding scientific correspondence between France and Prussia during this period. Even militarily sensitive information (for example in cartography) was freely exchanged by French and British mapmakers (...)." They were truly a caste above and separate from all others.

A certain Georges Cuvier, whom we met previously, survived unscathed throughout a five time dramatic regime change in France, which cost the lives of at least one million French soldiers and uncounted civilians who died through the horrors of the Revolution, Napoleonic and Civil wars. From the old kingdom through the entire revolution (when he was writing his articles as a modest 'citoyen Cuvier'), through the Napoleonic era (when he was made a Baron by Napoleon), the re–established monarchy and the constitutional monarchy after the 1830 July revolution, Cuvier was always aloft of the politics of the day, always honoured, always respected, always bestowed with medals, prizes and honourable memberships (at the end of his life by more than 80 academies and scientific societies).

Cuvier is not a singular case. The biographies of many important intellectual figures of the eighteenth and early nineteenth centuries are closely comparable. Only some odd–ones out had to flee their countries or died in poverty, alone and forgotten. Most enjoyed very successful and outwardly happy lives. In the case of Cuvier it was certainly overshadowed by the early deterioration of his eyesight, which ended his great studies on extant invertebrates and was a major driving force for him to concentrate on vertebrate palaeontology, and particularly the premature death of all his children.

What was so special about these men, that they were literally able to operate and co–operate even throughout times of war and revolution in a way nobody else could? How were they so unaffected by the major mishaps of humanity happening at the time? An aspect rather understudied with respect to the 'Republic of Letters' is its deep connection to freemasonry and other secret societies, as well as various other important subsets of society. Before you cry out 'conspiracy babble' hear the facts first and then judge by yourself. Many others were either members of the church, of the aristocracy, of the military or served in important political positions.

The book *"The invisible college: the Royal Society, Freemasonry and the Birth of Modern Science"* by Robert Lomas (2002) gives a glimpse to what extent freemasonry and science were entangled, particularly in Great Britain, since the founding days of the Royal Society. The same phenomenon can be seen in France and, to a lesser degree, in Germany and many other European countries.

I am not saying that being a Freemason makes you a bad person, or invalidates your scientific achievements. Outstanding geniuses like Johann Wolfgang von Goethe and Georges Washington were Freemasons. But being part of a hermetically closed, internationally operating society that includes presidents and princes certainly adds another layer to your social life and your potential political and economic influence that is not accessible to the public. How deeply masonic symbolism and influence is ingrained in the United States in particular since its inception is obvious if you just have a close look at the architecture of Washington D. C., the Statue of Liberty or any single dollar bill in existence.

Here is a list of eighteenth to early nineteenth century authors who contributed to vertebrate palaeontology whom I could positively identify as Freemasons, members of the church, military, aristocracy or who worked as politicians:

Freemasons:

Alleon–Dulac, Jean–Louis
Beckmann, Johannes
Born, Ignaz von
Buffon, Georges–Louis Marie Leclerc, Comte de (also an aristocrat via personal nobilitation)
Delamétherie, Jean–Claude
von Flurl, Mathias (member of the Illuminati order)
Griselini, Francesco
Lang, Friedrich Carl
Linné, Carl von
Martius, Ernst Wilhelm
Michaelis, Chrstian Friedrich
Picot de Lapeyrouse, Philippe–Isidore Baron (also an aristocrat via personal nobilitation and a politician)
Raspe, Rudolf Erich
Robinet, Jean–Baptiste–Réné
Roume de Saint–Laurent, Phillipe–Rose (also politician, French representative at Santo Domingo)
Soemmering, Samuel Thomas Ritter von (aristocrat via personal nobility)
von Schlotheim, Ernst Friedrich Baron (member of the Illuminati order, also aristocrat and politician)
von Schreber, Johann Christian Daniel Edler (also aristocrat)
Volney, Constantin–François Chassebouef Bosigrais Comte de (also an aristocrat and politician)
Werner, Gottlob Abraham

Wistar, Caspar
Zimmermann, Eberhard August Wilhelm von

Members of the church:

Annan, Robert (Presbyterian reverend)
Bertrand, Élie (Reformed pastor)
Beuth, Franz Martin (member of the jesuit order)
Bock, Friedrich Samuel (protestant military pastor)
Borlase, William (protestant priest)
Collin, Nicholas (protestant priest)
Dicquemare, Jaques François (Abbé)
Douglas, James (protestant cleric)
Falkner, Thomas (member of the jesuit order)
Fortis, Giovanni Battista (Abbé)
Goeze, Johann August Ephraim (protestant priest)
Graydon, Georges (protestant priest)
Jones, William (protestant cleric)
Kennedy, Ildephons (Benedictine monk)
Mann, Théodore–Augustin (Carthusian monk)
Meinecke, Johann Christoph (protestant cleric)
Nash, Treadway Russell (protestant cleric)
Palier, Johan Carel (protestant cleric)
Percy, Thomas (catholic bishop)
Pilkington, James (unitarian minister)
Pini, Ermenegildo (barnabite clegyman)
Polwhele, Richard (protestant clegyman)
de Ramatuelle, Thomas–Albin–Joseph d'Audibert (Abbé, also an aristocrat)
Silberschlag, Johann Esaias (protestant clergyman)
Soldani, Ambrogio (calmadolese monk)
Spallanzani, Lazzaro (member of the jesuit order)
Testa, Giovanni Domenico (Abbé)
Ure, David (minister of the Church of Scotland)
Volta, Giovanni Serafino (Abbé)

Aristocrats:

von Ansbach, Christian Friedrich Karl Alexander Markgraf
von Baczko, Ludwig
Barrington, Daines (4th son of the Viscount Barrington)
von Beroldingen, Franz Cölestin
d'Everlange de Vitry, Louis–Hyacinthe
Gazzola, Giovanni Battista Conte
von Huepsch, Johann Wilhelm Karl Adolph Freiherr
de Lamanon, Jean Honoré Robert de Paul
von Mellin, August Wilhelm (Graf)
de Razoumowsky, Gregoire Count
von Rochow, Friedrich Eberhard
Servières, Claude–Urbain de Retz Baron de (also an army officer)
Spadoni, Paolo
von Zach, Franz Xaver Freiherr (also an army officer)

Military:

Armstrong, John (major general and military engineer, General Quartermaster of Ireland, also an aristocrat)
Berkenhout, John (Prussian and British army officer, a British agent in America during the War of Independence)
Dillon, John Talbot (navy officer)
Edwards, Timothy (Colonel)
Garriga, José (Spanish army officer)
Imrie, Ninian (Scottish army officer)
Lasius, Georg Sigmund Otto (military engineer)
Nicola, Lewis (British and American officer)
Parsons, Samuel Holden (general of the Continental Army, also a politician)
Turner, Georges (officer of the Continental Army)

Politics:

Bossi, Luigi (agent of Napoleon in Turin, later archivary of the

King of Italy)

Brander, Gustavus (director of the Bank of England)

Buonamici, Giovanni Francesco (Italian diplomat)

Camper, Petrus (president of the state council of the Dutch republic)

Camper, Adriaan Gilles (Dutch statesman)

Cuvier, Georges (held many political positions, also aristocrat via personal nobility)

Defay–(Boutheroue), François Simon (French politician)

de La Coudreniere (Coudrenaire), (Henri Peyroux) (French politician)

Deluc, Jean–André (Swiss politician and diplomat)

Engel, Samuel (Swiss politician)

Faujas de Saint–Fond, Barthélemy (held political positions both under royal and revolutionary rule)

Geoffroy St–Hilaire, Étienne (Was the leading scientist who accompanied Napoleon to Egypt)

Giraud–Soulavie, Jean Louis (French diplomat, also a cleric)

Guyton de Morveau, Louis Bernard (French politician, also an aristocrat)

Jefferson, Thomas (many political postions, including the 3rd presidency of the USA)

Justi, Johann Heinrich Gottlob (many political positions)

Müller, Gottfried Adrian (Prussian politician)

Pownall, Thomas (British politician, Governor of Massachussets Bay)

Tilesius von Tilenau, Wilhelm Gottlieb (as politician in service of the Russian Czar. Aslo aristocrat via personal nobilitation)

de Saussure, Horace–Bénédict (active in swiss politics in his later years)

Schäffer, Jacob Jakob Gottlieb Christian von (court counselor of the prince of Thurn und Taxis, aristocrat via personal nobilitation)

There were other secret societies apart from the Freemasons

in existence and actively operating in the late eighteenth and nineteenth century, of which important figures in early vertebrate palaeontology were members.

Thomas Jefferson was, contrary to common belief, never provably a member of the Freemasons. He was, however, part of another secret society, so secret, in fact, that it has never become known what its initials "F. H. C." originally stood for (so it is usually humorously called the 'Flat Hat Club' in American literature). His son–in–law and his eldest grandson were Freemasons, though.

Johann Friedrich Blumenbach was a prominent member of the 'ZN'–order, a student order which had enlisted itself to the goals of enlightenment and was banned in 1778.

Martin Hinrich Carl Lichtenstein was a member of the quite influential 'Berliner Gesetzlose Gesellschaft', not really a secret society, but rather a loose circle of intellectual, political and military elite, which understood themselves as representing something like 'enlightenment with a national liberal touch'.

So, while easily available biographical data are not present for numerous authors (and therefore their affiliations cannot be definitely judged) in the cases of those that are, the list of guys interested in and working on vertebrate palaeontology in the late eighteenth and early nineteenth century with ties to 'secret societies', aristocracy, the church, politics and military is a long one.

That fact is in itself not really astounding. The 'ordinary Joe' in the eighteenth century probably had very little chances to get the academic training necessary to conduct any meaningful scientific work. Without a network of peers and sponsors you were completely lost, if you weren't a very wealthy man – publications in particular were costly, especially when they included good plates and illustrations, which are a must–have for palaeontological studies of any kind that are worth a dime.

Most people in eighteenth century Europe were still completely illiterate. If you were outside of one of those circles of 'high society', your chance to get anything meaningful done were close to zero. There are some cases where people from a modest background made a career as scientists, but these are unusual and rare. Most who worked on the subject of vertebrate palaeontology did so as a side–effect of their normal profession, and for almost zero of these authors it was a central topic of their research.

Georges Cuvier was one of the first scientists who devoted a very large part of his research curriculum exclusively to vertebrate palaeontology. This unique focus on the topic is probably what sets him apart most from all his contemporaries and predecessors, and the main reason why he was hailed far and wide as the 'godfather of palaeontology' up to the present day. And indeed it seems that, despite all his shortcomings, this at least can't be taken away from the man.

But what does this summary on the status of the early vertebrate palaeontology students have to do with our topic? Well, to me, it is just the best explanation available for why these scholars were so aloft of the political shenanigans of their day, usually at least.

Apart from being renowned scientists, they also were part of large international networks that did not care much about the quarrels between the 'normal plebs'. Such networks are found in the church and in 'secret societies' like the Freemasons, as well as the aristocracy, which, on an international scale, still held some kind of class–solidarity, which was particularly re–ignited after the gruesome events of the French Revolution, which literally cost many aristocrats their heads.

Those people therefore were not only members of the 'Republic of Letters', but more often than not they had additional affiliations to particular subsets and networks of society which, for a long time, had operated on a supra–national scale. There was a consensus among them about many things, which did not care much about national borders or quarrels, a sense of cosmopolitanism and internationalism, which also offered possibilities for mutual assistance and support in cases of crisis to which 'normal people' had little or no access. The Munich collection, containing Collini's famed *Pterodactylus*, being saved from dislocation to Paris just because of the personal friendship between von Moll and Haüy is a good case in relation to this.

The 'Republic of Letters' therefore, in summary, was just one of several closely intertwined 'Olympic rings' which provided prominent scholars who happened to work also on vertebrate palaeontology with a security and support system that operated on an international scale. The better you were connected within these multiple security systems, the better your chances to survive any kind of catastrophe happening around you at the time more or less unscathed, and to maybe even have influence on the course of events to a certain degree.

CHAPTER 12

What really happened – a forgotten world war, unlikely cataclysms and the 'Great Reset' of the nineteenth century.

The main thesis of the present book is that vertebrate palaeontology did not 'start' with Cuvier at all, and that Cuvier, in fact, wasn't even that important as we are made to believe, as practically everything fundamental that is found in his studies was already known, and widely so, before. Cuvier was made a 'poster–boy' for the 'new science'. For political reasons only, as the French Revolution, the rise and fall of Napoleon, and the monstrous catastrophes these events were for Europe (and far beyond) are very closely connected to this.

After the end of the Napoleonic wars and the never–ending series of cataclysms that had devastated both Europe and America from the time of roughly 1789 to 1820, the world had changed forever in ways that for us today are very difficult to grasp in all of their consequences. And Cuvier, a survivor of the catastrophe who escaped from all of it unharmed, was now the all–admired arch–priest of vertebrate palaeontology.

Very similar patterns can be found in practically all fields of the sciences through the same time–period, but I chose vertebrate palaeontology here for various reasons.

Firstly, it is the discipline which I have studied, worked in for decades and therefore know the best about.

Secondly, it was one of the most prominently emerging 'new sciences' of the nineteenth century, and is therefore of particular interest with respect to the cultural and civilizational 'reset' that took place during this time period.

Thirdly, it is a relatively small science. Even today, the scientific community of vertebrate palaeontologists (although it has risen from several dozen in the eighteenth century to several thousands now) is comparatively small and cosy.

It is therefore very much possible to analyse the history of this science throughout the late eighteenth and early nineteenth century, when only a few hundred authors were ever (officially) involved with the subject, in considerable detail and completeness. So it is ideally suited as an exemplary field of study on in what way, in what time frame and with what results the face of a science – and the world – was changed during this time.

We have already discussed the cataclysmic events which took place in Europe and the Americas during the latest eighteenth and early nineteenth century at some length in the preceding chapters. We are dealing with a time of constant war which, in permanently shifting theatres encompassing most of Europe, North– and South America as well as parts of Africa and Asia, took place pretty much on a world–wide scale (Mikaberidze 2020).

It involved almost all major military powers on Earth, only China and Japan, at that time still isolated from the rest of the world, being spared, although China almost entered war against Britain on one occasion, the trouble being caused by British attempts to occupy the Portuguese colony of Macao in 1808 (Wood 1940). China was, however, dealing with a series of massive internal conflicts itself after 1795, which resulted in numerous long–lasting domestic wars during that timeframe, including the Miao Revolt (1795–1806), the White Lotus rebellion (1796–1805) and the Eight Trigram Sect uprising (1813) (Elleman 2005).

Furthermore, a series of natural catastrophes of devastating effect took place during that timeframe. The most outstanding one was probably the rarely discussed Tambora explosion at Sumbawa Island, Indonesia, the largest volcanic eruption ever witnessed by past–stone–age men, which took place in 1815.

Oppenheimer (2003) summarizes some of its effects: "This [eruption] formed a global sulfate aerosol veil in the stratosphere, which resulted in pronounced climate perturbations. Anomalously cold weather hit the northeastern USA, maritime provinces of Canada, and Europe the following year. 1816 came to be known as the 'Year without a summer' in these regions. Crop failures were widespread, and the eruption has been implicated in accelerated emigration from New England, and widespread outbreaks of epidemic typhus."

Right after the end of the Napoleonic wars, America and Europe were now struck with what Post (1977) called "The last great subsistence crisis to affect the Western world". The years immediately following the Napoleonic period were characterized by widespread famine, raging pandemics, mass migration, financial crises, panics and collapses. No one has ever counted the victims.

In both Europe and North America, a devastating series of fires destroyed entire cities, particularly in the United States, where these urban fires were probably the greatest scourge of the country throughout the nineteenth century. Boston, New York,

New Orleans, Pittsburgh, Washington, Chicago and about 100 other major cities in the US were hit – sometimes several times – by devastating urban fires.

And then there were the little known series of New Madrid earthquakes of 1811–12, the worst in the history of the United States, about which Johnston & Schweig (1996) had the following to say: "Continental North America's greatest earthquake sequence struck on the western frontier of the United States. The frontier was not then California but the valley of the continent's greatest river, the Mississippi, and the sequence was the New Madrid earthquakes of the winter of 1811–1812. Their described impacts on the land and the river were so dramatic as to produce widespread modern disbelief. However, geological, geophysical, and historical research, carried out mostly in the past two decades, has verified much in the historical accounts." And this unprecedented catastrophe struck the country right before the second American–British war, which lasted from 1812 to 1814/15.

And these few examples are just the tip of the iceberg. Fascinating times, and interestingly the times in which modern vertebrate palaeontology, as well as practically our entire modern world, emerged.

What happened during this time and afterwards can only be described, even if we completely follow historical text–book knowledge, as a civilization reset. Imagine huge global powers of today, like the USA or the People's Republic of China just disappearing from the map, with new superpowers that weren't even there before suddenly emerging, let's say, in of all places, Africa?

That was exactly what happened. Dominant superpowers for uncounted centuries like the Spanish Empire and the Holy Roman Empire were gone for good, whereas new ones like the United States of America, which just had not existed before, emerged.

Despite the massive efforts of 'restoration' in England and Continental Europe, the ideals and ideas of the American and French Revolution had taken a big foothold and could never ever be erased again. The status quo ante became forever unattainable, and the heydays of the industrial revolution with their massive effects on society, politics and economics were just around the corner.

Even Africa was headed for disaster. After being for the longest time a secondary priority for the European colonial powers, it moved into the spotlight after most European colonies in the 'New World' had been lost. By the end of the nineteenth century, the entire African continent (except Ethiopia and, arguably, Liberia) was under European colonial rule.

It is out of this cataclysmic change of the world that most of what we know as modern science emerged, and vertebrate palaeontology is no exception. Only that, as we have hopefully learned by now, its true history is much distorted by myths and legends, by accidental loss and deliberate memory–holing. And lies.

Specimens, collections, even manuscripts, records and scientific works of great importance gone. The entire history streamlined ad usum delphini and focussing on just a few key–figures and key–events. We will probably never know the full truth. Too much has disappeared under the cultural layers, the debris of ages gone by. History has truly been stolen.

But I hope that this little essay has at least helped the reader to catch a glimpse of what comes a bit closer to the truth than the things you read in most textbooks and popular accounts. I also hope that some readers may become encouraged to do their own research into the lost and forgotten history of early palaeontology, or maybe other aspects of the history of science, or history in general.

There is so much to be learned, and our view is largely blocked by so many white spots, that we are still barely scratching the surface.

Appendix 1: Taxa of fossil vertebrates named in the eighteenth century

Scrotum humanum Brookes, 1768

Rhinoceros lenenensis Pallas, 1773 (nomen oblitum)

Elephas americanus Kerr, 1792 American Mastodon

Ursus spelaus Rosenmüller, 1794 (Cave bear)

Megatherium americanum Cuvier, 1796

Pycnodus apodus (Volta, 1796)

Monopterus gigas Volta, 1796

Pegasus volans Volta, 1796

Ramphosus rastrum (Volta, 1796)

Ductor vestenae (Volta, 1796)

Vomeropsis triurus (Volta, 1796)

Mene rhombea (Volta, 1796)

Exellia velifer (Volta, 1796)

Archaephippus asper (Volta, 1796)

Eoplatax papilio (Volta, 1796)

Carangodes bicornis (Volta, 1796)

Eocottus veronensis (Volta, 1796)

Pygaeus bolcanus (Volta, 1796)

Blochius longirostris Volta, 1796

Xiphopterus falcatus Volta, 1796

Megalonyx Jefferson, 1797 (*jeffersonii* Harlan, 1825)

Coelodonta antiquitatis (Blumenbach, 1799) Wooly rhinoceros

Megaloceros giganteus (Blumenbach, 1799) Irish elk

Elephas primigenius (Blumenbach, 1799) Wooly mammoth

Elephas mammonteus Cuvier, 1799 (junior synonym of *Elephas primigenius* Blumenbach, 1799)

Mammut Blumenbach, 1799 (for *Elephas americanus*)

References

Abilgaard, P. C. (1796): Kort Beretning om det Kongelige Naturalcabinet i Madrid, med en Beskrivelse over et gigantisk Skelet af et nyt ubekiendt Dyr, som er opgravet i Peru og bevares i dette Museum. Nye Samling af det Kongelige Danske Videnskabernes Selskabs Skrifter, **1799**(5): 402-414, 1 lám. Johan Rudolph Thiele, Kiobenhavn (Published serapartely in 1796).

Agassiz, L. (1833-1844): Recherches sur les Poissons fossiles. 5 volumes, atlas. Neuchâtel (Petitpierre).

Allain, R. (2001): Redescription de *Streptospondylus altdorfensis*, le dinosaure théropode de Cuvier, du Jurassique de Normandie. — Geodiversitas, **23**(3): 349-367.

[Anonymous] (1757): [Further accounts of fossils.] — Gentleman's Magazine, **27**: 122– 123.

Barthel, K. W. (1978): Solnhofen: ein Blick in die Erdgeschichte. 393 pp.; Thun (Ott).

Bauhin, J. (1598): Historiae fontis et balnei admirabilis Bollensis liber quartus. 222 pp.; Mömpelgard (Foillet).

Bianucci, G. & Landini, W. (2007): Fossil history. — In: Reproductive Biology and Phylogeny of Cetacea: Whales, Porpoises and Dolphins,7: 35-93.

Bilfinger, G. B. (1743): Varia in fasciculos collecta. 314 pp.; Stuttgart (Erhardt).

Blainville, H. M. D. de (1818): Poissons fossiles. In: Nouveau Dictionnaire d'Histoire naturelle appliqué aux arts, à l'agriculture, à l'économie rurale et domestique, à la médecine, etc Vol. XXVII, pp. 310-396, Paris (Deterville).

Blumenbach, J. F. (1799): Handbuch der Naturgeschichte. 6th ed., Göttingen xvi + 708 pp., 2 pls.

Blumenbach, J. F. (1807): Handbuch der Naturgeschichte., 8th ed., Göttingen (H. Dieterich).

Boyd, J. P. (1958): The *Megalonyx*, the *Megatherium*, and Thomas Jefferson's Lapse of Memory. — Proceedings of the American

Philosophical Society, **102**(5): 420-435.

Breyne,Johann Philip (& Wolochowicz, M.) (1741): A Letter from John Phil. Breyne, M. D. F. R. S. to Sir Hans Sloane, Bart. Pres. R. S. with Observations, and a Description of Some Mammoth's Bones Dug up in Siberia, Proving Them to Have Belonged to Elephants. Philosophical Transactions of the Royal Society of London, **40**: 124–138.

Brookes, R (1763): The Natural History of Waters, Earths, Stones, Fossils, and Minerals, with their Virtues, Properties, and Medicinal Uses: To Which is added, The Method in which Linnaeus has treated these Subjects, Vol. V (London: Newberry).

Bryson, B. (2010): A short history of nearly everything. 560 pp.; New York (Broadway Books).

Buckland, W. (1824): Notice on the *Megalosaurus*, or great fossil lizard of Stonesfield. — Transactions of the Geological Society of London. ser. 2, **1**: 390-396.

Buckland, W. (1829): On the discovery of a new species of pterodactyle in the Lias at Lyme Regis. — Transactions of the Geological Society of London, ser. 2, **3**: 217–222.

Burtin, F.-X. (1790): Les revolutions générales, qu'a subiès la surface de la terre, et sur l'ancienneté de notre globe. — Verhandelingen uitgegeeven door Teyler's Tweede Genootschap, **VIII**: 242 pp. + Dutch transl.

Camper, A. G. (1790): Lettre de A.G. Camper-G. Cuvier sur le ossements fossiles de la montagne de St. Pierre - Maestricht. — Bulletin de la Societé Philosophique, Fructidor **8**: 306-315.

Camper, A. G. (1791): Lettre de A.G. Camper-G. Cuvier sur le ossements fossiles de la montagne de St. Pierre - Maestricht. — J. Phys. Vend. **9**: 109-117.

Camper, A. G. (1800): Lettre de A.G. Camper- G. Cuvier sur le ossements fossiles de la montagne de St. Pierre - Maestricht. — Journal de Physique, de Chimie, d'Histoire Naturelle et des Arts, **51**: 278-291.

Catesby, M. (1734-1747): The natural history of Carolina, Florida and the Bahama Islands. Vol. II. London.

Ciampini, G. C. (1688): Comparazione delle osse fossili di Vitorchiano, nel Viterbese, con quelle di uno scheletro d'elefante. Roma.

Cohen, C. (2002): The fate of the mamooth: fossils, myth and history. XXXIV & 297 pp.; Chicago and London (Chicago University Press).

Cohen, C. & Wakefield, A. (2008): Gottfried Wilhelm Leibnitz, Protogaea, Translated and edited by, Claudine Cohen and Andre Wakefield. 216 pp.; Chicago and London (Chicago University Press).

Collini, C. A. (1784): Sur quelques zoolithes du cabinet d'Histoire Naturelle de S.A.S.E. Palatine et de Baviere, a Manheim. — Acta Acadamiae Theodoro Palatinae, Mannheim, Pars Physica, **5**: 58–103.

Cooper, W. (1828): Further Discovery of Fossil Bones in Georgia: and remarks on their identity with those of the *Megatherium* of Paraguay. — Annals of the Lyceum of Natural History of New York, **2**(1): 267-270.

Collareta, A.; Collareta, M.; Berta, A.; & Bianucci, G. (2021): On Leonardo and a fossil whale: a reappraisal with implications for the early history of palaeontology. — Historical Biology, **33**(10): 2289–2298.

Conybeare, W. D. (1824): On the Discovery of an almost perfect Skeleton of the *Plesiosaurus*. — Transactions of the Geological Society, London, (2)**1**: 381-389.

Cuvier, G. (1796a): Notice sur le squelette d'une très-grande espèce de quadrupède inconnue jusqu'à present, trouvé au Paraguay, et déposé au cabinet d'histoire naturelle de Madrid. — Magasin encyclopédique, ou Journal des sciences, des lettres et des arts (Paris), **1796** (1): 303-310, 2 pls.; **1796** (2): 227-228.

Cuvier, G. (1796b): Memoire sur les especes d'Elephans tant vivantes que fossiles, lu a la seance publique de FInstitut national le 15 germinal, an IV [April 4, 1796]. Magasin encyclopédique; 1796; 2. annee 3: 440-445.

Cuvier, G (1798): Extrait d'un ouvrage sur les espèces de quadrupèdes dont on a trouvé les ossemens dans l'intérieur de la terre. — Bulletin de la Societé Philomatique, Paris, **18**: 137–139 [Ser. I] 1 [Pt. 2] also in: Magasin encyclopédique, 4e année 3 (1798) 145–150; and in: Journal de Physique, Chimie, et d'Histoire Naturelle de l'Institut National 47 (1798) 313–317. [6]

Cuvier, G. (1799): Mémoir sur les espèces d'éléphans vivants et fossiles. — Mémoires de l'Institut national de sceinces et arts. Sciences mathématiques et physiques, **II**: 1-22, pls. ii-vi.

Cuvier, G (1800a): Extrait d'un ouvrage sur les espèces de quadrupèdes dont on a trouvé les ossemens dans l'intérieur de la terre, Baudouin, Paris, 1800.

Cuvier, G. (1800b): A quantity of bones found in the rocks in the environs of Honfleur, by the late Abbé Bachelet. — Philosophical Magazine VIII: 290

Cuvier, G. (1801): Extrait d'un ouvrage sur les espèces de quadrupe`des dont on a trouve´ les ossemens dans l'inteérieur de la terre. — Journal de Physique, de Chemie et d'Histoire Naturelle, **52**: 253– 267.

Cuvier, G. (1804): Sur le *Megatherium*. Autre animal de la famille des Paresseux, mais de la taille du Rhinocéros, dont un squelette fossile presque complet est conserv6 au cabinet royal d'histoire naturlle à Madrid. — Annales du Muséum national d'Histoire Naturelle, Paris, **5**: 376-387.

Cuvier, G. (1808): Sur le grand animal fossile des carrières de Maestricht. — Annales du Muséum national d'Histoire Naturelle, Paris, **12**: 145-176.

Cuvier, G. (1809a): Sur quelques quadrupèdes ovipares fossiles conservés dans des schistes calcaires. — Annales du Muséum d'Histoire Naturelle, Paris, **13**: 401-437.

Cuvier, G. (1809b): Mémoire sur le squelette fossile d'un reptile Volant des environs d'Aichstedt, que quelques naturalistes ont pris pour un oiseau, et don't nous formons un genre de Sauriens, sous le nom de Petro-Dactyle. — Annales du Muséum national d'Histoire Naturelle, Paris, **13**: 424–437.

Cuvier, G (1812): Recherches sur les Ossemens Fossiles des Quadrupèdes, 4 vols, Deterville, Paris.

Cuvier, G. 1819: (*Pterodactylus longirostris*). — Isis von Oken, **2**: 1128, 1788.

Cuvier, G. (1821-1824): 2nd ed, Recherches sur les Ossemens Fossiles des Quadrupèdes, 5 vols, G. Dufour & E. d'Ocagne, Paris, 1821–1824.

Cuvier, G. (1834-1836): 4th ed., Recherches sur les Ossemens Fossiles des Quadrupèdes, 5 vols, E. d'Ocagne, Paris.

Darwin, C. (1838): A sketch of the deposits containing extinct Mammalia in the neighbourhood of the Plata. — Proceedings of the Geological Society of London, **2**: 542-544.

Daston, L. (1991): The Ideal and Reality of the Republic of Letters in the Enlightenment. — Science in context, **4**(2): 367-386.

Daubenton, L. J.-M. (1764): Mémoire sur des os et des dents remarquables par leur grandeur. — Mémoires de'l Académie Royale des Sciences, Paris, an. 1762: 206-229.

Davis, L. E. (2012): Mary Anning: Princess of Palaeontology and Geological Lioness. — The Compass: Earth Science Journal of Sigma Gamma Epsilon, **84**(1): 56-88.

Delair, J. B. & Sarjeant, W. A. S. (1975): The aerliest discoveries of dinosaurs. — Isis, **66**: 4-25.

Delair J. B. & Sarjeant, W. A. S. (2002): The earliest discoveries of dinosaurs: the records re-examined. — Proceedings of the Geological Association, 113: 185–197.

Dicquemare, J. F. (1776): Ostéolithes. — Journale de Physique, de Chimie, d' Histoire naturelle et des Arts, **7**: 406-414.

Elleman, B. A. (2005): Modern Chinese Warfare, 1795-1989. IX & 362 pp.; London & New York (Routledge).

Eskildsen, K. R. (2004): How Germany left the republic of letters. — Journal of the History of Ideas, **65**: 421-432.

Esper, J. F. (1774): Description des zoolithes nouvellement découvertes d'animaux quadrupedes inconnus et des cavernes qui les renferment de même que de plusieurs autres grottes remarquables qui se trouvent dans le Margraviat de Baireuth au delà des monts. Fol., Nuremberg 121 pp., 14 pls.

Etheridge, K. (2014): Leonardo and the whale. — In: Fiorani F (ed.): Leonardo da Vinci: between art and science NEH summer institute essays. accessed 2014 May 3. https://faculty.virginia.edu/Fiorani/NEH-Institute/essays/etheridge.

Etheridge, K. (2019): Leonardo and the whale. — In: Moffatt C, Taglialagamba S (eds.): Leonardo da Vinci – nature and architecture. 89–106; Leiden (Brill).

Evans, M. (2010): The roles played by museums, collections and collectors in the early history of reptile palaeontology. — Geological Society, London, Special Publications, **343**(1): 5-29.

Falkner, T. (1774): A description of Patagonia and the adjoining parts of South America. 144 pp., maps ; Hereford and London.

Faujas de Saint-Fond, B. (1799): Histoire naturelle de la montagne de Saint-Pierre de Maestricht. 4[quarto], Paris 263 pp., 54 pls.

Flower, W. H. (1898): Essays on Museums. — London (Macmillan).

Fraas, E. (1891): Die Ichthyosaurier der süddeutschen Trias- und Juraablagerungen. 81 pp.; Tübingen (Laupp).

Fumaroli, M. (1988): The republic of letters. — Diogenes, **36**: 129-152.

Garriga, J. (1796): Descripción del esqueleto de un quadrúpedo muy corpulento y raro, que se conserva en el Real Gabinete de Historia Natural de Madrid. Fol., Madrid xvii + 20 pp., 5 pls.

Gaudant, J. (2010): Histoire d'une brève controverse : Wilhelm Ernst Tentzel (1659-1707) et l'éléphant fossile découvert en 1695 à Burg Tonna, près de Gotha (Allemagne). — Travaux du Comité français d'Histoire de la Géologie, 2010, 3ème série (tome 24), pp.117-130.

Gaudant, J. (2011): Bréve histoire de la collcetion de Gazola de poissons fossiles éocènes de Monte Bolca (Italie) conservée au Museum nationale d'histoire naturelle. — Geodiversitas, **33**: 637-647.

Ges(s)ner, J. (1758): Tractatus physicus de petrificatis in duas partes distinctus, quarum prior agit de petrificatorum differentiis, et eorum varia origine; altera vere de petrificatorum variis originibus, praecipuarumque telluris mutationum testibus. 136 pp.; Leiden.

Goldfuss, G. A. (1831): Beiträge zur Kenntnis verschiedener Reptilien der Vorwelt. I. Reptilien aus dem lithographischen Schiefer. — Nova Acta Physico-Medica Academiae Caesaerae Leopoldino-Carolinae Naturae Curiosum, **15**: 61–128.

Guettard, L. E. (1756): Suite du memoire dans lequel on compare le Canada à la Suisse, par rapport a ses mineraux. — Mémoires de l'Academie des Sciences, Année 1752: 323-360,

Gunther, R. T. (1925): Early Science in Oxford. The Biological Collections, 3 Volumes. (the Author).

Gunther, R.T. (1945): Early Science in Oxford. Life and Letters of Edward Lhuyd, vol, 14. Oxford (the Author) [Reprinted 1967 London (Dawsons of Pall Mall].

Halstead, L. B. (1970): *Scrotum humanum* Brookes 1763-The First Named Dinosaur. — Journal of Insignificant Research, **5**:14-15.

Halstead L. B. & Sarjeant, W. A. S. (1993): *Scrotum humanum* Brookes – the earliest name for a dinosaur. — Modern Geology, **18**: 221–224.

Hall, B. K. (2002): Palaeontology and evolutionary developmental biology: a science of the nineteenth and twenty-first centuries. — Palaeontology, **45**(4): 647-669.

Hay, O. P. (1902): Bibliography and catalogue of the fossil Vertebrata of North America. — U. S. Geological Survey, Bulletin No. **179**: 1-868.

Hay, O. P. (1929-1930): Second bibliography and catalogue of the fossil Vertebrata of North America. — Carnegie Institution Washington, Publication No. **390**. 2 volumes.

Hiemer. E. F. (1724): Caput Medusae utpote novum diluvii universalis monumentum: detectum in agro Würtembergico et brevi dissertatiuncula epistolari expositum. 40 pp.; Stuttgart (Rösslin).

Howe, S. R.; Sharpe, T. & Torrens, H. S. (1981): Ichthyosaurs: A History of Fossil 'Sea-dragons'. — National Museum of Wales, Cardiff

Howlett, E.; Kennedy, W. J.; Powell, H. P. & Torrens, H. (2019): New light on the history of *Megalosaurus*, the great lizard of Stonesfield. — Archives of Natural History, 44: 82-102.

Hunter, W. (1768): Observations on the Bones, Commonly Supposed to Be Elephants Bones, Which Have Been Found Near the River Ohio in America: By William Hunter, M. D. F. R. S. — Philosophical Transactions of the Royal Society of London, **58**: 34–45.

Jaeger, G. F. von (1824): De ichthyosauri sive proteosauri fossilis speciminibus in agro Bollens in Würtembergia repertis. 14 pp.; Stuttgart (Cotta).

Jaeger, G. F. von (1828): Über die fossilen Reptilien, welche in Würtemberg aufgefunden worden sind. 46 pp.; Stuttgart (Metzler).

Johnston, A. C. & Schweig, E. S. (1996): The enigma of the New Madrid earthquakes of 1811–1812. — Annual Review of Earth and Planetary Sciences, 24(1): 339-384.

Jones, R. T. (2011): Peter Simon Pallas, Siberia, and the European Republic of Letters. — İstoriko-biologiçeskie issledovaniya, **3**: 55-67.

Karg, J. M. (1805): Über den Steinbruch zu Oehningen bei Stein am Rhein und dessen Petrefakten. — Denkschriften der vaterländischen Gesellschaft der Ärzte und Naturforscher Schwabens, **I**: 1-74.

Kempe, S.; Rosendahl, W. & Döppes, D. (2005): The Making of the Cave Bear – Die wissenschaftliche Entdeckung des „Ursus spelaeus". — Mitteilungen der Kommission für Quartärforschung der Österreichischen Akademie der Wissenschaften, **14**: 89–106.

Kerr, R. (1792): The animal kingdom, or zoological system of the celebrated Sir Charles Linnaeus; class I, Mammalia, being a translation of that part of the Systema Naturae published by Prof. Gmelin. xii + 28 + 644 + 8 pp., London.

Keyßler, J. G. (1740-41): Neueste Reisen durch Deutschland, Böhmen, Ungarn, die Schweiz, Italien und Lothringen: worinnen der Zustand und das Merkwürdigste dieser Länder beschrieben, und vermittelst der Natürlichen, Gelehrten und Politischen Geschichte, der Mechanik, Maler- Bau- und Bildhauerkunst, Münzen und Alterthümer, wie auch mit verschiedenen Kupfern erläutert wird. 1556 pp., 2 volumes; Hannover (N. Förster).

Lamanon, J. H. R. De P. (1783): Lettre relative à l'ornitholithe de Montmartre. — Journal de physique, de chimie, d'histoire naturelle et des arts, **XXII**: 309-313.

Langenmantel, H. A. (1689): De ossibus Elephantum. — Miscellanea curiosa sive ephemeridum medico-physicarum Germanicarum Academiae Caesareo-Leopoldinae Naturae Curiosorum, Dec. II, anno 7 (1687): p. 446-447.

Lennier, G. (1887): Etudes paleontologiques. Description des fossiles du Cap de la Heve. — Bulletin de la Societe Geologique de Normandie, **12**: 17-98.

Lingham-Soliar, T. (1995): Anatomy and functional morphology of the largest marine reptile known, *Mosasaurus hoffmanni* (Mosasauridae, Reptilia) from the Upper Cretaceous, Upper Maastrichtian of the Netherlands. — Philosophical Transactions of the Royal Society of London. Series B, **347**(1320): 155-180.

Lister, A. (2018): Darwin's Fossils: The Collection that Shaped the Theory of Evolution. 160 pp.; Washington D. C. (Smithsonian Institution).

Lomas, R. (2002): The Invisible College: The Royal Society, Freemasonry and the Birth of Modern Science. 384 pp.;

London (Headline Book Publishing).

Lhuyd, E. (1699): Lithophylacii Britannici Ichnographia, sive, lapidium aliorumquefoss ilium Britannicorum singularifigura insignium. London (Gleditsch and Weidmann).

Mantell, G. A. (1824): Description of some fossil vegetables of the Tilgate Forest in Sussex. — Transactions of the Geological Society, London, **1**: 421–424.

Mantell, G. A. (1829): A tabular arrangement of the organic remains of the country of Sussex. — Transactions of the geological Society, London, **2**(3): 201-216.

Martill, D. M. (2010): The early history of pterosaur discovery in Great Britain. — Geological Society, London, Special Publications, **343**: 287-311.

Mayor, A. (2007): Fossil legends of the first Americans. 446 pp.; Princeton (Princeton Univerity Press).

Meyer, H. von. (1832): Palaeologica, zur Geschichte der Erde und ihrer Geschöpfe. 560 pp.; Frankfurt am M. (S. Schmerber).

Mikaberidze, A. (2020): The Napoleonic Wars: A Global History. XXIII + 936 pp.; Oxford (Oxford University Press).

Mullen, A. (1682): An anatomical account of the elephant accidentally burnt in Dublin on Fryday, June 17 in the year 1681, sent in a letter to Sir Will. Petty, fellow of the Royal Society : together with a relation of new anatomical observations in the eyes of animals, communicated in another letter to the honorable R. Boyle... fellow of the same Society. Printed for Sam. Smith, London, 42 p

Murr, C. G. von (1774): Herrn Dr. Hunters Anmerkungen über die sogenannten Elephantenknochen, welche am Ohiostrome in America gefunden wurden. — Der Naturforscher, **3**: 237-239.

Nichols, J. B. (1795-1800): The history and antiquities of the County of Leicester. Fol., 4 vols.; London (J. Nichols).

Noè, L. F.; Gómez-Pérez, M., & Nicholls, R. (2019): Mary Anning, Alfred Nicholson Leeds and Steve Etches. Comparing the three most important UK 'amateur'fossil collectors and their collections. — Proceedings of the Geologists' Association, **130**(3-4): 366-389.

Oken, L. [On *Ornithocephalus*] — Isis von Oken, **2**: 246–253.

Orbigny, A. d' (1842): Voyage dans l'Amérique Méridionale (le Brésil, la République Orientale de l'Uruguay, la République Argentine, la Patagonie, la République du Chili, la République de Bolivia, la République du Pérou), exécuté pendant les années 1826, 1827, 1828, 1829, 1830, 1831, 1832 et 1833. Tome Troisième, 4° Partie: Paléontologie. Paris et Strasbourg (P. Bertrand et V. Levrault).

Owen, R. (1839): Description of a tooth and part of the skeleton of the *Glyptodon*, a large quadruped of the edentate order, to which belongs the tessellated bony armour figured by Mr. Clift in his memoir on the remains of the Megatherium, brought to England by Sir Woodbine Parish, F.G.S. — Proceedings of the Geological Society of London, **3**: 108-113.

Oppenheimer, C. (2003): Climatic, environmental and human consequences of the largest known historic eruption: Tambora volcano (Indonesia) 1815. — Progress in physical geography, 27(2): 230-259.

Osborn, H. F. (1935): Thomas Jefferson as a paleontologist. — Science, **82**(2136): 533-538.

Padian, K. & Wild, R. (1992): Stuides of Liassic Pterosauria, 1. The holotype and referred specimens of the Liassic pterosaur *Dorygnathus banthensis* (Theodori) in the Petrefaktensammlung Banz, northern Bavaria. — Palaeontographica A, **225**: 59-77.

Padian, K. (2008): The early Jurassic pterosaur Dorygnathus banthensis (Theodori, 1830). — Special Papers in Paleontology, **80**: 1-64.

Pander, C. H. von & d'Alton, J. S. E. (1821): Vergleichende Osteologie, I, 1: Das Riesen-Faulthier, *Bradypus giganteus*, abgebildet, beschreiben, und mit den verwandten

Geschlechtern verglichen. Bonn (Weber).

Pieters, F. F.; Rompen, P. G.; Jagt, J. W., & Bardet, N. (2012): A new look at Faujas de Saint-Fond's fantastic story on the provenance and acquisition of the type specimen of Mosasaurus hoffmanni Mantell, 1829. — Bulletin de la Société géologique de France, **183**(1): 55-65.

Piñeiro, J. M. L. (1988): Juan Bautista Bru (1740-1799) and the Description of the Genus *Megatherium*. — Journal of the History of Biology, 21(1): 147-163.

Platt, J. (1759): An account of the fossile thigh-bone of a large animal, dug up at Stonesfield, near Woodstock, in Oxfordshire. — Philosophical Transactions of the Royal Society London, **L**(2): 524-527.

Plot, R. (1677): The Natural History of Oxfordshire, Being an Essay toward the Natural History of England (1st ed., Oxford).

Podgorny, I. (2012): Fossil dealers, the practices of comparative anatomy and British diplomacy in Latin America, 1820-1840. — The British Journal for the History of Science, 46(4): 647-674.

Post, J.D. (1977): The last great subsistence crisis the Western World. Baltimore MD: The Johns Hopkins University Press, 240 pp.

Quenstedt, F. A. (1856-1858): Der Jura. 1066 pp.; Tübingen (Laupp).

de Razoumowsky, G. C. (1790): Observations propres à prover que toute la Suisse grèseuse et toute la plaine peu sinueuse du Cercle de Baviere, doivent leur origine aux eaux douces Lacustres. — Histoire et mémoires de la Société des Sciences Physiques de Lausanne, **III**: 204-236, map.

Rebstock, J. M. (1720): Göttliche Erkenntnuss aus wunderbaren figurierten Steinen, so in der Gegend des Boller Bades, Zeller Bades und Zeller Stabes in der Erden, Wasser, auch in dem Schiefer und den Steinen gefunden wurden. Ulm.

Rieppel, O. (2021): The first ever described dinosaur bone fragment in Robinet's philosophy of nature (1768). — Historical Biology.

Rosenmüller, J. C. (1794): Quaedam de ossibus fossilibus animalis cuiusdam, historiam eius et cognitionem accuratiorem illustrantia. 4[quarto], Diss., Leipzig 34 pp., pl.

Roume de Saint-Laurent, P.-R. (1795-1796): (1799 in Romer et al.): Squelette fossile trouvé sur les bords de la Plata. — Bulletin de la Societé Philomatique, Paris I(2): 96-97.

Russell, D.A. (1967): Systematics and morphology of American Mosasaurs. — Bulletin of the Peabody Museum of natural History, 23: 1-237.

Scaramucci, J. B. (1697): Meditationes familiares ad Antonium Magliabechium de sceleto elephantino a celeberrimo Wilhelmo Ernesto Tentzelio ; ubi quoque testaceorum petrificationes defendutur, et aliqua subterranea phenomena examini subiiciuntur. 24 pp.; Urbini (Litteris Leonardi)

Schávelzon, D. & Arenas, P. (1992): Los inicios de la Paleontología en Argentina. — Todo es Historia, **295**: 37-49.

Scheuchzer, J. J. (1726): Homo Diluvii Testis Et Theoskopos Publicae Suksitisi Expositus. 24 pp.; Zürich (J. H. Bürgklin).

Scilla, A. (1670): La vana speculazione disigannata dal senso: lettra responsiva circa i corpi marini, che petrificati si trovano in varii luoghi terrestri. 206 pp.; Napoli (A. Colicchia).

Simpson, G. G. (1942): The beginnings of vertebrate paleontology in North America. — Proceedings of the American Philosophical Society, **86**(1): 130-188.

Smith, J. C. (1993): Georges Cuvier: an annotated bibliography of his published works. Smithsonian Institution Press, Washington DC.

Soemmerring, S. T. von (1812): Über einen *Ornithocephalus*. — Denkschriften der Akademie der Wissenschaften München, Mathematisch–Physikalische Klasse, **3**: 89–158.

Soemmerring, S. T. von (1817): Über einen *Ornithocephalus brevirostris* der Vorwelt. — Denkschriften der Akademie der Wissenschaften München, Mathematisch–Physikalische Klasse, **6**: 89– 104.

Stukely, W. (1719): An Account of the impression of the almost entire sceleton of a large animal in a very hard stone, lately presented to the Royal Society, from Nottinghamshire. — Philosophical Transactions of the Royal Society, **30**: 963–968.

Taquet, P. & Padian, K. (2004): The earliest restoration of a pterosaur and the philosophical origins of Cuvier's *Ossemes fossiles*. — Comptes Renues Palevol, **3**: 157-175.

Tentzel, W. E. (1696): Epistola de sceleto Elephantino Tonnae nuper effosso, ad virum toto orbe celeberrimum Antonium Magliabechium, Serenissimi Magni Hetruriae Ducis Bibliothecarium et consiliarium. 16 pp.; Gotha (Literis Reyherianis).

Tentzel, W. E. (1698): Epistola de sceleto Elephantino Tonnae nuper effosso, ad virum toto orbe celeberrimum Antonium Magliabechium. – Philosophical Transactions of the Royal Society of London, **19**: 757-776

Theodori, C. von (1830): Knochen vom Pterodactylus aus der Lias-Formation von Banz. Frorieps Notizen für Natur- und Heilkunde, **632**: 101-103.

Theodori, C. von (1831): Über die Knochen vom Genus Pterodactylus aus der Lias-Formation von Banz. – Isis, **3**: 276-281.

Theodori, C. von (1852): Über Pterodactylusknochen im Lias von Banz. — Berichte des Naturforschenden Vereins Bamberg, **1**: 17-44.

Tischlinger, H. (2020): Der „Collini-*Pterodactylus*" - eine Ikone der Flugsaurier-Forschung. — Archaeopteryx, 36: 16-31.

Tischlinger, H. (2023): Eichstätter Fossilien im fürstbischöflichen Hochstiftskalender von 1759. — Archaeopteryx, 38: 68-76.

Thomson, K. S. (2008): The legacy of the Mastodon: The golden age of fossils in America. Yale University Press.

Torrens, H. S. (1979): Geological communication in the Bath area in the last half of the eighteenth century. — In: Jordanova, L. J. & Porter, R. S. (eds) Images of the Earth: Essays in the History of the Environmental Sciences. BSHS Monographs, 1. British Society for the History of Science, Chalfont St Giles, 215–247.

Van Miert, D. (2014): What was the Republic of Letters? A brief introduction to a long history. — Groniek, (204/5), 129-284.

Van Marum, M. (1790): Beschrijving der beenderen van den kop van eenen visch, gevonden in den St. Pietersberg by Maastricht en geplaatst in Teyler's Museum. — Verhandelingen uitgegeeven door Teyler's Tweede Genootschap, **VIII**: 383-389.

Vega, Garilasco de la (1609): Comentarios reales de los Incas. Lisboa (Crasbeeck).

Vogel, R. A. (1776): Practisches Mineralsystem. 2d ed., Leipzig 12 + 582 + 10 pp.

Volta, G. S. (1796): Ittiolitologia veronese del Museo Bozziano ora annesso a quello del Conte Giovambattista Gazola e di altri Gabinetti di fossili veronesi. Fol., Verona lii + cccxxiii pp., 76 pls. (1796-1809).

Wellnhofer P. (2008): A short history of pterosaur research. — Zitteliana, **B28**: 7–18.

Weishampel, D. B. & Young, L. (1996): Dinosaurs of the East Coast. Johns Hopskins University Press.

Werner, G. A. (1774): Von den äußerlichen Kennzeichen der Fossilien. Lepizig (Crusius).

Wood, H. J. (1940): England, China, and the Napoleonic Wars. — Pacific Historical Review, **9**(2): 139-156.

Woodward, J. (1728): A Catalogue of the Additional Extraneous English Fossils, London.

Ziegler, B. (1986): Der Schwäbische Lindwurm – Funde aus der Urzeit. 171 pp. , Stuttgart (K. Theiss).